To permit maximum use of the outpouring of historical articles on science which accompanied the celebration of the American bicentennial, George W. Black, Jr., Science Librarian, Southern Illinois University, Carbondale, systematically compiled all bicentennial references to science and technology which appeared in over fifteen hundred journals in 1976.

The resulting bibliography contains some 1,065 entries arranged under headings taken from the Wilson index or other standard lists, such as L. C. subject headings and Index medicus. Indexes to journal titles, to authors, and to proper names cited in the articles enable the user to find any item of information easily and quickly. Complete paging of the articles indexed is given, together with cross-referencing to other articles by the same author or authors included in the bibliography.

A useful tool for readers seeking information on subjects ranging from the practice of acupuncture in 1836 to Continental army camp followers.

American Science and Technology

A Bicentennial Bibliography

George W. Black, Jr.

Southern Illinois University Press
Carbondale and Edwardsville
Feffer & Simons, Inc.
London and Amsterdam

Library of Congress Cataloging in Publication Data

Black, George W.
  American science and technology.

  Includes indexes.
  1. Science—United States—History—Bibliography.
2. Technology—United States—History—Bibliography.
I. Title.
Z7405.H6B55 [Q125]   016.509'73   78-15820
ISBN 0-8093-0898-3

Contents

Preface

Criteria for inclusion

The Bibliography is an outgrowth of observations
on the compiler's part that the passage through our
national bicentennial celebration stimulated an outpouring of historical articles commemorating the
event. Individual issues of over fifteen hundred
journals (including all journals listed in Biologi-
cal and agricultural index, Applied science and
technology index, and science-techology journals
listed in the Readers' guide, Business periodicals
index and Education index) were scanned for articles related to the bicentennial. To keep the
project manageable only 1976 issues of the journals
were checked. When leads to articles published
earlier or later were found they were followed up,
and if appropriate, they were included. In the
course of the preparation of the Bibliography I became aware of other historical articles published
during the year. For example, 1976 also was the
centennial anniversary of the invention of the telephone by Alexander Graham Bell. I decided to in-
clude nonbicentennially related material if it 1)
pertained to the history of American science or
technology and 2) discussed developments which oc-
curred prior to the twentieth century. As a final

check, the major science-technology abstracting or
indexing services were scanned to assure the most
comprehensive coverage possible. Some relevant
articles were not included due to errors on my part,
late-arriving interlibrary loans, incomplete or
slow listings by abstract or index services, and
the like.

Intended audience

I visualize the Bibliography as serving two purposes. First, as a unique, if incomplete, historical record of the scientific community's response
to the event of our national bicentennial. Second,
as a useful tool for both the school and public librarians which should enable them to quickly iden-
tify biographical or historical articles pertaining
to the history of American science or technology.
Complete paging of the articles is included together with the number of reference cited by the
authors.

Subject headings and abbreviations

Since the work was primarily intended for use by
school and public librarians, wherever possible, I
used headings taken from Wilson index (especially
Applied science and technology index and Biologi-
cal and agricultural index). When I was unable to
find appropriate headings in these indexes I used
other standard lists such as L.C. subject headings
and Index medicus. I also attempted to stay close
to Wilson abbreviations for journal titles. If any
ambiguities exist, I believe they could easily be
resolved by consulting the Journal Title Index of

the book.

Credits

I am indebted to many people who offered advice
and encouragement to me in the course of the Bibliography preparation. I would especially like to
recognize Mrs. Marybelle Melvin, former Assistant
Science Librarian at Morris Library; Mrs. Marge
Conway, Science Division Chief Library Clerk; my
children, Kathy, Mary, and Jim who spent many hours
proofreading early drafts of the manuscript; and
especially my wife, Katherine, who took time from
her own busy schedule to proofread the final draft.

I also wish to express my appreciation to the
Office of Research Development and Administration,
Southern Illinois University, Carbondale, which pro-
vided the vital funds I needed to complete the
Bibliography and prepare the manuscript for pub-
lication.

George W. Black, Jr.
Carbondale, Illinois
June 1978

American Science and Technology

Bibliography

1   ABBE, CLEVELAND
        Hetherington, Norriss S. "Cleveland Abbe and
        a view of science in mid-nineteenth century
        America." <u>ANN</u> <u>SCI</u> 331(1) Jan '76, 31-49
        (120 ref.)
        ABBOTT, CHARLES CONRAD. See MAN--PREHISTORIC--
        NEW JERSEY 525
2   ACCIDENTS--CHRONOLOGY
        McCall, Brenda. "1776-1976: days to remem-
        ber." <u>PROF</u> <u>SAFETY</u> 21(1) Jan '76, 27-30; (2)
        Feb '76, 11-15; (3) Mar '76, 45-51; (4) Apr
        '76, 40-43; (5) May '76, 42-45; (6) Jun '76,
        44-47; (7) Jul '76, 45-49; (8) Aug '76,
        48-52; (9) Sep '76, 46-49; (10) Oct '76,
        51-54; (11) Nov '76, 50-54; (12) Dec '76,
        49-52
3   ACUPUNCTURE--19th CENTURY
        Greenwood, Ronald D. "Acupuncture in the
        United States, 1836." <u>SOUTH</u> <u>CAROL</u> <u>MED</u> <u>ASSN</u> <u>J</u>
        72(5) May '76, 182-83 (1 ref.)
4   AGGREGATES
        "Lightweight aggregate, a late developing
        industry." <u>BRICK</u> <u>CLAY</u> <u>REC</u> 169 (1) Jul '76,
        33

An overview of American land policy. Paul W.
Gates; Research and education in American
agriculture. Paul E. Waggoner; Plant intro-
ductions. Knowles A. Ryerson; Changes in
animal science. T.C. Byerly; The pioneer
farmer: A view over three centuries. Gilbert
C. Fite; Midwest agriculture. Changing with
technology. Hiram M. Drache; Agriculture two
hundred years from now. Don Paarlberg

8   AGRICULTURE
Heady, Earl O. "The agriculture of the U.S."
SCI AMER 235 (3) Sep '76, 106-08, 110, 115,
118, 121, 122-23, 126-27 (3 ref.)

9   -----
"The miracle of U.S. agri-power." FARM CHEM
139(10) Oct '76, 15-20, 22, 24

10 -----
Rasmussen, Wayne D. "Two centuries of Ameri-
can farming and exports." FOR AGRI 14(41)
Oct 11, '76, 2-5, 11

11 -----
Shea, Kevin P. "American agriculture." ENVIR-
ONMENT 18(8) Oct '76, 28-38 (39 ref.)

12 AGRICULTURE--FORECASTS
Logsdon, Gene. "Leave us alone and we'll
produce the food." FARM J 100 (7) Jun/Jul
76, 16-17, 20

13 -----
Seim, Dick. "Tomorrow's farmer. . . here
today." FARM J 100 (7) Jun/Jul '76, 13-15

14 AGRICULTURE--GEORGIA
Plummer, Gayther L. "Mulberries to soybeans:
changing vegetation patterns." GA ACAD SCI

BULL 34(4) Sep '76, 182-91 (14 ref.)

15   AGRICULTURE--PRE-COLUMBIAN

"The first American farmers." FARM J 100 (1)
Jan '76, 53-55, 60

16   AIR CONDITIONING

"Air conditioning: instant comfort." EXXON
CHEM MAG 9(3) '76, 14-15

17   -----

Macleod, E. "The history of room air condit-
ioners." ASHRAE J 18(7) Jul '76, 41-42

-----. See also CARRIER, WILLIS H. 135

18   AIR PUMPS--18th CENTURY

Smoluk, George R. "Two-cylinder pump replaces
bellows in blast furnace." DESIGN N 31(13)
Jul 4, '76, 38-39

19   AIRSHIPS

Robinson, Douglas H. "Dr. August Greth and
the first airship flight in the United
States." AMER AVIATION HIST SOC J 21(2) Sum
'76, 84-91 (15 ref.)

See also BALLOONS--18th CENTURY 81

ALCOHOLISM. See TEMPERANCE 892

20   ALCOHOLISM--18th CENTURY

Rush, Benjamin. "An inquiry into the effects
of ardent spirits on the human body and
mind." ALCOHOL HEALTH RES WORLD Sum '76, 7-9

21   ALCOHOLISM--PENNSYLVANIA

Glaser, Frederick B. "Alcoholism in Pennsy-
lvania--a Bicentennial perspective." PENN
MED 79(7) Jul '76, 70-80 (72 ref.)

ALEXANDERSON, ERNST F.W. See STEINMETZ,
CHARLES P. 844

22 ALGORITHMS

Walbesser, Henry H. "Algorithms and the Bicentennial." MATH TEACHER 69(5) May '76, 414-18 (5 ref.)

23 ALTER, DAVID

Hodge, Edwin S. "David Alter and other spectroscopists in Western Pennsylvania." APP OPTICS 15(7) Jul '76, 1722-25 (15 ref.)

ALUMINUM. See HALL, CHARLES MARTIN 735

24 AMERICA--CIVILIZATION

De Lauer, Marjel. "The bi-millennium of the southwest." ARIZONA HIGHWAYS 52(1) Jan '76, 2-9

25 AMERICA--DISCOVERY AND EXPLORATION

"Across the Appalachians." NAT PARKS & CON MAG 50(5) May '76, 13-15

26 AMERICA--DISCOVERY AND EXPLORATION--MEDICAL ASPECTS

Lopez, Cesar A. "Medical notes on Columbus's first trip to America." AMER MED ASSN J 236 (14) Oct 4, '76, 1598-99 (15 ref.)

27 AMERICAN CERAMIC SOCIETY--GLASS DIVISION

Pincus, Alexis G. "The Glass Division's role." GLASS IND 57 (4) Apr '76, 16-18

28 AMERICAN CHEMICAL SOCIETY

Reese, Kenneth M. "American Chemical Society: The first 100 years." CHEM & ENG N 54(15) Apr 6, '76, 22-31 (2 ref.)

29 AMERICAN DERMATOLOGICAL ASSOCIATION

Szymanski, Frederick J. "Centennial history of the American Dermatological Association-- 1876 to 1976." ARCH DERM 112 (Spec. Issue) Nov 29, '76, 1651-53

30   AMERICAN GYNECOLOGICAL SOCIETY
          Everett, Houston S., and E. Stewart Taylor.
          "The history of the American Gynecological
          Society and the scientific contributions of
          its Fellows." AMER J OBSTET GYNECOL 126(7)
          Dec 1, '76, 908-19 (32 ref.)
31   AMERICAN INSTITUTE OF ELECTRICAL ENGINEERS
          McMahon, A. Michael. "Corporate technology:
          The social origins of the American Institute
          of Electrical Engineers." IEEE PROC 64(9)
          Sep '76, 1383-90 (35 ref.)
32   AMERICAN PHYSICAL THERAPY ASSOCIATION
          "The beginning of the Association." PHYS
          THER 56(1) Jan '76, 50-62 (13 ref.)
33   AMERICAN PHYSICAL THERAPY ASSOCIATION--OFFICERS
          "Presidents of the APTA (1921-1976), 63-65;
          Executive Directors, 66: Editors of the
          Journal (1921-1976) 67." PHYS THER 56(1)
          Jan '76
34   AMERICAN PSYCHIATRIC ASSOCIATION
          "The Original Thirteen." HOSP COMMUNITY
          PSYCHIATRY 27(7) Jul '76, 464-67
35   AMERICAN SOCIETY FOR INFORMATION SCIENCE
          Schultz, Claire K. "ASIS: Notes on its
          founding and development."
          AMER SOC INFORM SCI BULL 2(8) Mar '76, 49-51
      AMERICAN TELEPHONE AND TELEGRAPH COMPANY. See
          TELEPHONE 882
36   AMERICAN VETERINARY MEDICAL ASSOCIATION
          Freeman, Arthur. "A brief history of the
          AVMA." AMER VET MED ASSN J 169(1) Jul 1, '76,
          120-26 (1 ref.)

AMISH. See POPULATION GENETICS 725

37   AMMANN, O.H.

    Rigoni, Donald L., Jr. "O. H. Ammann: The view from the bridge." CIVIL ENG-ASCE 46(7) Jul '76, 77

    ANDERSON, MARY CATHERINE. See MEDICAL CARE-- NEW JERSEY--19th CENTURY 542

    ANESTHESIA. See LONG, CRAWFORD WILLIAMSON 506

38   ANESTHESIA--19th CENTURY

    "Anesthesia--American discovery." MED ASSN STATE ALA J 45 (11) May '76, 18. See also MED SOC NJ J 73(4) Apr '76, 330B

39   -----

    "Anesthesia for surgery--an American discovery." KAN MED SOC J 77(7) Jul '76, 347-51 (4 ref.)

40   ANESTHESIOLOGY--NEW YORK

    Betcher, Albert M. "History of anesthesiology in New York." NY STATE J MED 76(7) Jul '76, 1165-70 (52 ref.)

41   ANTIQUES IN INTERIOR DECORATION

    Ward, Gerald W. R. "Elegance in Revolutionary America." CRAFT HORIZONS 36(2) Apr '76, 52-55 (6 ref.)

42   ARCHITECTURE

    "Highlights of American architecture, 1776-1976." AIA J 65(7) Jul '76, 88-158

43   -----

    Peck, Robert. "Federal architecture, U.S. property keep off." PROG ARCH 57(7) Jul '76, 46-51

44   -----

    Tower, Edward M. "American architectural

heritage." <u>MIL</u> <u>ENG</u> 68(443) May–June '76,
207-9

45   ARCHITECTURE--AMERICAN MIDWEST
Miller, Brian. "Buildings of the tall grass
prairie." <u>AIA</u> <u>J</u> 65(7) Jul '76, 172-75

46   ARCHITECTURE BY BLACKS
Dozier, Richard. "The Black architectural experience in America." AIA J 65(7) Jul '76,
162-64, 166, 168

47   ASTRONAUTICS
"Toward man's dream of worlds unseen; 1950-
1976." <u>ASTRONOMY</u> 4(7) Jul '76, 98-107
-----. See also MARS (PLANET) 528-29

48   ASTRONOMICAL PHOTOGRAPHY
Bell, Trudy E. "History of astrophotography."
<u>ASTRONOMY</u> 4(7) Jul '76, 66-79

49   ASTRONOMY
Bell, Trudy E. "The beginnings of American
astronomy." <u>SKY</u> <u>TELE</u> 52(1) Jul '76, 26-30

50   -----
Osterbrock, Donald E. "The California-Wiscon-
sin Axis in American astronomy." <u>SKY</u> <u>TELE</u>
51(1) Jan '76, 9-14; (2) Feb '76, 91-97

51   -----
"Practical men, practical astronomy: 1776-
1825." ASTRONOMY 4(7) Jul '76, 34-43
-----. See also ABBE, CLEVELAND 1

52   ASTRONOMY--18th CENTURY
Galatola, Anthony. "The beginnings of Ameri-
can astronomy." <u>SCI</u> <u>&</u> <u>CHILD</u> 13(4) Jan '76,
26-28

53   ASTRONOMY--19th CENTURY
      "Astronomy comes of age: 1825-1840." ASTRO-
      NOMY 4(7) Jul '76, 44-49
      -----. See also RUTHERFURD, LEWIS MORRIS 784
54   ASTRONOMY--20th CENTURY
      "The universe unfolds: 1900-1950." ASTRONOMY
      4(7) Jul '76, 86-89, 95
55   ASTRONOMY--CHRONOLOGY
      Mendillo, Michael, David DeVorkin, and Rich-
      ard Berendzen. "History of American astrono-
      my--a chronological perspective." ASTRONOMY
      4(7) Jul '76, 20-21
56   ASTRONOMY--PRE-COLUMBIAN
      Chamberlain, Von Del. "Prehistoric American
      astronomy (c. 1054 A.D.)." ASTRONOMY 4(7)
      Jul '76, 10-11, 15-19
57   -----
      Hicks, Robert D., III. "Astronomy in the
      ancient Americas." SKY TELE 51(6) Jun '76,
      372-77
58   -----
      Libassi, Paul T. "Observatories without tele-
      scopes." SCIENCES 16(2) Mar/Apr '76, 11-15
59   -----
      Phillips, Henry J. "Skylore of indigenous
      Americans." ASTRONOMY 4(7) Jul '76, 12-14
60   ASTROPHYSICS--19th CENTURY
      "Astrophysics is born: 1840-1900." ASTRONOMY
      4(7) Jul '76, 50-63
61   ATOMIC POWER PLANTS
      Smiley, Seymour H., Malcolm L. Ernst, George
      Sege, and Robert T. Jaske. "Feeding the glut-
      ton." IEEE SPECTRUM 13(7) Jul '76, 74-83

-----. See also TRADE ASSOCIATIONS--AUTOMOTIVE
INDUSTRY 904

72   AUTOMOTIVE INDUSTRY--BIOGRAPHY
"The movers and the shakers." MOTOR AGE 95
(6) Jun '76, 73-74, 77, 78, 81, 82

73   -----
"A salute to the men behind the progress."
AUTOMOT IND 155(1) Jul 1, '76, 102-8

74   AUTOMOTIVE INDUSTRY--CHRONOLOGY
"Automobiles that built the nation, history-
tracing wall chart 1892-1976." AUTOMOT IND
155(1) Jul 1, '76, insert between 84-85

75   AUTOMOTIVE INDUSTRY--FORECASTS
"Challenge and the progress." AUTOMOT IND
155(1) Jul 1, '76, 125-28

76   AUTOMOTIVE INDUSTRY--LAWS AND REGULATIONS
"Government's ominous role." AUTOMOT IND
155(1) Jul 1, '76, 115-18

77   -----
"Government's ominous role." MOTOR AGE 95(6)
Jun '76, 61-65

78   AUTOMOTIVE SERVICE EQUIPMENT
Marinucci, Dan. "Automotive service equipment has
come a long way, baby." MOTOR AGE 95(6) Jun '76,
34-37, 39

79   AVIAN MEDICINE
Witter, J. Frank. "The history of avian medicine in the United States. I. Before the big
changes." AVIAN DIS 20(4) Oct-Dec '76, 621-
30

80   AXES
Smoluk, George R. "Steel edge and curved han-
dle produce superior axe." DESIGN N 31(13)

Jul 4, '76, 30-31

BACON, CYRUS, JR. See GETTYSBURG, BATTLE of
1863 364

BALLOONS. See AIRSHIPS 19; BLANCHARD, JEAN
PAUL 105-06; UNITED STATES CIVIL WAR 945

81   BALLOONS--18th CENTURY
Robinson, Douglas H. "The first aerial voyage
in America." AMER AVIATION HIST SOC J 21(1)
Spr '76, 3

82   BANNEKER, BENJAMIN
Bedini, Silvio A. "Benjamin Banneker, The
first Black man of science." SCI & CHILD 13
(4) Jan '76, 19-21

83   -----
"Benjamin Banneker (1731-1806)." ASTRONOMY
4(7) Jul '76, 38

84   -----
Mulcrone, T. F. "Benjamin Banneker, pioneer
Negro mathematician." MATH TEACHER 69(2) Feb
'76, 155-60 (11 ref.)

85   BARNEY, JOSHUA
Wilkinson, Dave. "The legend of Joshua Barney." OCEANS 9(6) Nov-Dec '76, 4-13

86   BARTRAM, JOHN
"John Bartram--America's first botanist."
DESIGN N 31(13) Jul 4, '76, 37

87   -----
Precup, A. V. "John Bartram: 1699-1777."
BIOSCIENCE 26(5) May '76, 359

88   BECHET de ROCHE FONTAINE, ETIENNE
Buzzaird, Raleigh B. "Washington's last
chief engineer." MIL ENG 68(446) Nov-Dec '76,
452-55

89   BEEF
       Black, Roe, and Warren Kester. "How beef be-
       came No. 1 on America's table." FARM J 100
       (4) Mar '76, G-1, G-4, G-8
90   BEERS, CLIFFORD
       "Clifford Beers: A man for a cause." HOSP
       COMMUNITY PSYCHIATRY 27(7) Jul '76, 493-94
91   BEES
       Ambrose, John T., and William G. Lord. "A
       Bicentennial salute to the honeybee." GLEAN-
       INGS 104(7) Jul '76, 248-49
92   -----
       Oertel, Everett. "Bicentennial bees, early
       records of honey bees in the eastern United
       States." AMER BEE J 116(2) Feb '76, 70-71;
       (3) Mar '76, 114, 128; (4) Apr '76, 156-57;
       (5) May '76, 214-15; (6) Jun '76, 260-61,
       290 (37 ref.)
       -----. See also HUBER, FRANCIS 408
93   BEES--19th CENTURY
       Oertel, Everett. "Old reports about honey-
       bees and beekeeping in United States docu-
       ments." GLEANINGS 104(1) Jan '76, 10, 35
       (1 ref.); (2) Feb '76, 51-52; (3) Mar '76,
       106-7
94   BEES--MASSACHUSETTS--17th CENTURY
       Murray, Lee. "The first bees in Massachuset-
       ts." AMER BEE J 116(7) Jul '76, 336
95   BEES--NEW YORK
       Wixsom, Grace T. "The story of one of Ameri-
       ca's first beekeeping families." AMER BEE J
       116(6) Jun '76, 262-63

96   BELL, ALEXANDER GRAHAM
      "Alexander Graham Bell." POST OFF ELECTR ENG
      J 69(1) Apr '76, 2
97   -----
      "Alexander Graham Bell--More than telephone
      inventor." NAT WOOL GROW 66(6) Jun '76, 11
98   -----
      Berger, J. Joel. "Alexander Graham Bell: in-
      ventor of electric speech." SCI TEACH 43(5)
      May '76, 36-39 (5 ref.)
99   -----
      Crowther, J. G. "Invention that set the world
      a-talking." NEW SCI 69(991) 11 Mar '76, 574-
      76
100   -----
      Hounshell, David A. "Bell and Gray: contrasts
      in style, politics, and etiquette." IEEE
      PROC 64(9) Sep '76, 1305-14 (71 ref.)
      -----. See also TELEPHONE 879
101   BELL, ALEXANDER GRAHAM--PICTORIAL WORKS
      Brannan, Beverly W., with Patricia T. Thomp-
      son. "Alexander Graham Bell, a photograph
      album." LIB CONG Q J 34(2) Apr '77, 73-96
      (12 ref.)
      BENNETT, JESSE. See CESAREAN SECTION 144
      BETHLEHEM, PENNSYLVANIA. See WATERWORKS--
      BETHLEHEM, PENNSYLVANIA--18th CENTURY 1028
      BILLINGS, JOHN SHAW. See SANITATION--SURVEYS
      796
102   BIOMEDICAL RESEARCH
      Burger, Alfred. "Biomedical sciences: the
      past 100 years." CHEM & ENG N 54(15) Apr 6,
      '76, 146-52, 157-60, 162 (35 ref.)

BIRKHOFF, GEORGE D. See DYNAMICAL SYSTEMS 242

103 BLACK FAMILIES

Scott, Roland B., and Michael R. Winston. "The health and welfare of the black family in the United States, a historical and institutional analysis." AM J DIS CHILD 130(7) Jul '76, 704-7 (19 ref.)

104 BLACK; JOSEPH

Pratt, Herbert T. "Samuel L. Mitchill's evaluation of the lectures of Joseph Black." J CHEM ED 53(12) Dec '76, 745-46 (23 ref.)

105 BLANCHARD, JEAN PIERRE

Blanchard, Jean Pierre. "My forty-fifth ascension: the first aerial voyage in America." TRANSLOG 7(3) Mar '76, 7-8, 19-21

106 -----

Hirschfeld, Fritz. "Of Blanchard's balloon and Washington's passport." MECH ENG 98(8) Aug '76, 20-21

BLIGH, WILLIAM. See FOOD 308

107 BON HOMME RICHARD

Drewry, John M. "Sea search for history: Project Bon Homme Richard." SEA TECH 17(5) May '76, 18-19

108 BONDS--WATER DISTRICTS

Fielding, Elizabeth. "Future of bonds tied to cities' future." WATER WASTES ENG 13(7) Jul '76, 31-32

BOONE, DANIEL. See AMERICA--DISCOVERY AND EXPLORATION 25

109 BOOTH, JAMES CURTIS

Stock, John T. "James Curtis Booth and his balance." J CHEM ED 53(8) Aug '76, 497-98

(4 ref.)

BOSTON--SIEGE, 1775-1776--MEDICAL ASPECTS. See
JACKSON, HALL 455

110 BOSTON--SIEGE, 1775-1776--WEATHER
Ludlum, David M. "The weather of American
independence--2: The siege and evacuation of
Boston 1775-76." WEATHERWISE 27(4) Aug '76,
162-68 (30 ref.)

111 BOTANY--NOMENCLATURE
Crowley, Webster R., Jr. "Early America in
plant names." MORTON ARBOR Q 12(2) Sum '76,
12-29; (3) Aug '76, 40-44

BLAST FURNACES. See AIR PUMPS--18th CENTURY 18

112 BOTANY, MEDICAL
Der Marderosian, Ara and Mukund S. Yelvigi.
"Sources of drugs used by Amerindians and
Colonists during the Colonial Period and
thereafter." AMER J PHARM 148(4) Jul-Aug '76,
118-20

113 BOWDITCH, NATHANIEL
"Nathaniel Bowditch (1773-1838)." ASTRONOMY
4(7) Jul '76, 39

114 -----
Rink, Paul E. "Nathaniel Bowditch--the practical navigator." NAVIGATOR 22(1) Win '75,
16-24

BRADFORD, WILLIAM. See LAISSEZ-FAIRE 477

115 BRADLEY, SAMUEL BEACH
Atwater, Edward C. "Samuel Beach Bradley,
M.D., 1796-1880, a rural practitioner." NY
STATE J MED 76(10) Oct '76, 1883-88

BRASHEAR, JOHN A. See ALTER, DAVID 23

BREADFRUIT. See FOOD 308

116 BRICKMAKING
        "The development of machinery made brick-
        making an industry." BRICK CLAY REC 169(1)
        Jul '76, 30-31
117 BRICKS
        "Designers of America wanted permanence:
        they chose brick." BRICK CLAY REC 169(1)
        Jul '76, 20-21
118 BRIDGER, JIM
        Hewitt, Bob, "Jim Bridger--mountain man."
        WEST HORSE 41(2) Feb '76, 14-15, 91-94
119 BRIDGES
        "America's greatest suspension bridges."
        CIVIL ENG--ASCE 46(7) Jul '76, 78-79
120 -----
        Watson, Sara Ruth. "Some historic bridges of
        the United States." J PROF ACTIV (ASCE PROC)
        101(EI3) Jul '75, 383-90 (6 ref.)
        -----. See also AMMANN, O. H. 37
121 BRIDGES--18th CENTURY
        Knowlton, Nancy B. "Covered bridge spans the
        Schuykill." DESIGN N 31(13) Jul 4, '76, 32-
        33
122 BROWN, LUCIUS POLK
        Wolfe, Margaret R. "The making of a state
        scientist-the preparation of Lucius Polk
        Brown, 1867-1908." TENN ACAD SCI J 51(1)
        Jan '76, 2-6 (10 ref.)
        BROWN, MOSES. See TEXTILE INDUSTRY 894
123 BUILDINGS, BRICK--PICTORIAL WORKS
        "Early brick buildings." BRICK CLAY REC 169
        (1) Jul '76, 24-25

124 BUNKER HILL, BATTLE OF--MILITARY ENGINEERING
     Mason, John H., and Walter F. Mackie.
     "Bunker Hill." MIL ENG 68(443) May-Jun '76,
     166-69
125 BURGOYNE'S INVASION, 1777--WEATHER
     Ludlum, David M. "The weather of indepen-
     dence--6: Burgoyne's northern campaign."
     WEATHERWISE 29(5) Oct '76, 236-40; (6) Dec
     '76, 288-90 (7 ref.)
126 BURNET, WILLIAM
     "A Latin inaugural by Dr. William Burnet,
     1767." MED SOC NJ J 73(5) May '76, 456
127 BYRD, WILLIAM A.
     Hagler, Carl. "A country surgeon." ILL MED J
     149(5) May '76, 477
128 CANALS
     McNown, John S. "Canals in America." SCI
     AMER 235(1) Jul '76, 116-24 (4 ref.)
     -----. See also AUGUSTA (GEORGIA) CANAL 62;
     ERIE CANAL 288; GORGAS, WILLIAM CRAWFORD 368
129 CANCER
     Shimkin, Michael B. "What do we know about
     cancer?" WEST J MED 125(6) Dec '76, 509-12
130 -----
     Shubik, Phillippe, and David B. Clayson.
     "Environmental cancer and chemical agents
     1976." ANN INTERN MED 85(1) Jul '76, 120-22
131 -----
     Van Scott, Eugene J. "Cancer perspectives--
     1876 to 1976, benefits and hazards." ARCH
     DERM 112(Spec. Issue) Nov 29, '76, 1666-67
     -----. See also RUSH, BENJAMIN 781

132   CANCER NURSING--FORECASTS
      Koons, Shirlee B. "The future of cancer
      nursing." RN 39(8) Aug '76, 23, 27-28, 32, 34
133   CARDIOLOGY
      Bishop, Louis F. "Cardiology as a specialty."
      NY STATE J MED 76(7) Jul '76, 1170-74
134   -----
      Burchell, Howard B. "Selected vignettes rela-
      tive to the Bicentennial and American cardi-
      ology." MOD CONCEPTS CARDIOVASC DIS 45(1)
      Jan '76, 71-76 (42 ref.)
      -----. See also HEART DISEASES 385
      CARDIOLOGY - RHODE ISLAND. See Fulton, Frank
      TAYLOR 349
135   CARRIER, WILLIS H.
      Kessler, Ellen. "Air conditioning and Dr.
      Carrier." ASHRAE J 18(7) Jul '76, 45-46
136   CATESBY, MARK
      Bell, Joseph. "Mark who?, Catesby . . . .
      Mark Catesby, a British naturalist in Co-
      lonial America." ANIMAL KINGDOM 79(5) Oct/
      Nov '76, 12-24
137   CATTERTON, DYLER
      Kirkwood, Tom. "Counties first physician
      settled near Pinkstaff." ILL MED J 149(2)
      Feb '76, 164-65
138   CATTLE
      McMurray, Patty. "The Bicentennial cow."
      HOARD'S DAIRY 121(13) Jul 10, '76, 795, 801
139   CERAMIC INDUSTRY--FORECASTS
      "And what next? A decade, a century." CER
      IND 107(1) Jul '76, 33-37, 40

140 CERAMIC MATERIALS--ELECTRIC PROPERTIES
      "Sophistication in ceramics." CER IND 107(1)
      Jul '76, 30-31, 44
141 CERAMICS
      Mueller, James I. "Our ceramic heritage."
      AMER CER SOC BULL 55(7) Jul '76, 677-79
142 -----
      "Our ceramic heritage." AMER CER SOC BULL
      55(1) Jan '76, 159-60
      -----. See also U.S. CENTENNIAL CELEBRATIONS,
      ETC. 943
143 CERAMICS--PICTORIAL WORKS
      "Early American ceramics at Yale." CER MO 24
      (5) May '76, 21-23
144 CESAREAN SECTION
      King, Arthur G. "The legend of Jessee Ben-
      net's 1794 Cesarean Section." BULL HIST MED
      50(2) Sum '76, 242-50 (29 ref.)
      CHAPIN, HENRY DWIGHT. See INFANTS--MORTALITY
      429
      CHEMICAL APPARATUS. See BOOTH, JAMES CURTIS
      109
145 CHEMICAL EDUCATION
      Newell, Lyman C. "Chemical education in Amer-
      ica from the earliest days to 1820." J CHEM
      ED 53(7) Jul '76, 402-4 (15 ref.)
146 CHEMICAL EDUCATION--18th CENTURY
      Adcock, Louis H. "Early American chemistry
      teachers." CHEMISTRY 48(10) Nov '75 15-16
      (5 ref.)
      -----. See also RUSH, BENJAMIN 777
147 CHEMICAL INDUSTRY
      Henahan, John F. "200 years of American

chemicals." CHEM WEEK 118(7) Feb 18, '76,
25-40, 45-60

148 -----
Pigford, Robert L. "Chemical technology: the
past 100 years." CHEM & ENG N 54(15) Apr 6,
'76, 190-200, 202-3 (26 ref.)

149 CHEMICAL INDUSTRY--18th CENTURY
Adcock, Louis H. "Manufacturing chemists in
the American Revolution." CHEMISTRY 48(9)
Oct '75, 10-11

150 CHEMISTRY
White, J. Edmund. "Chemistry and the U.S.A."
J CHEM ED 53(12) Dec '76, 738-40

151 CHEMISTRY--18th CENTURY
Adcock, Louis H. "Founding fathers and other
patriots." CHEMISTRY 49(2) Mar '76, 21-23
(6 ref.)

152 -----
Beer, John J. "The chemistry of the Founding
Fathers." J CHEM ED 53(7) Jul '76, 405-8
(25 ref.)

153 -----
May, Ira P., and William E. Wort. "Chemistry
in Colonial America." CHEMISTRY 49(6) Jul-
Aug '76, 6-7 (7 ref.)

154 CHEMISTRY--19th CENTURY
Bernheim, Robert A. "Chemistry in 1876: the
way it was." CHEM & ENG N 54(15) Apr 6, '76,
38-42, 47-51 (31 ref.)

155 -----
Ihde, Aaron J. "European tradition in nine-
teenth century American chemistry." J CHEM
ED 53(12) Dec '76, 741-44 (22 ref.)

156   CHEMISTRY--CALIFORNIA
        Norberg, Arthur L. "Chemistry in California."
        CHEM & ENG N 54(36) Aug 30, '76, 26-36 (19
        ref.)
157   CHEMISTRY--CHRONOLOGY
        "Chronology of important chemical and rela-
        ted events since 1876." CHEM & ENG N 54(15)
        Apr 6, '76, 91-92
158   CHEMISTRY, ANALYTIC
        Belcher, R. "American Bicentennial, 1776-
        1976, two hundred years of Anglo-American
        analytical chemistry." ANAL CHIM ACTA 86(1)
        Oct '76, 1-11 (18 ref.). See also CHEM BRIT
        12(12) Dec '76, 387-89
159   -----
        Egan, Harold. "200 years of Anglo-American
        analytical chemistry, applied aspect." CHEM
        BRIT 12(12) Dec '76, 389-90
160   -----
        Ewing, Galen W. "Analytical chemistry: the
        past 100 years." CHEM & ENG N 54(15) Apr 6,
        '76, 128-34, 139-42 (43 ref.)
161   CHEMISTRY, INORGANIC
        Zubieta, Jon A., and Jerold J. Zuckerman.
        "Inorganic chemistry: the past 100 years."
        CHEM & ENG N 54(15) Apr 6, '76, 64-79
162   CHEMISTRY, ORGANIC
        Tarbell, D. Stanley. "Organic chemistry: the
        past 100 years." CHEM & ENG N 54(15) Apr 6,
        '76, 110-18, 121-23 (19 ref.)
163   CHEMISTRY, PHYSICAL
        Eyring, Henry. "Physical chemistry: the past
        100 years." CHEM & ENG N 54(15) Apr 6, '76,

88-90, 93-94, 99, 101, 103-4

164 CHEMISTRY, POLYMER

Mark, Herman F. "Polymer chemistry: the past 100 years." CHEM & ENG N 54(15) Apr 6, '76, 176-89

165 CHEMISTS--GEORGIA--19th CENTURY

Whitehead, T.H. "Georgia's nineteenth century chemists." GA ACAD SCI BULL 34(4) Sep '76, 200-203 (5 ref.)

166 CHILDBIRTH--19th CENTURY

McGrellis, Nyra M. "Labor and delivery 120 years ago." JOGN 5(3) May-Jun '76, 56-58

167 CHILDREN

Newman, Lucile F. "A Bicentennial view of childrearing: notes of an anthropologist." RHODE ISLAND MED J 59(5) May '76, 221-23, 241-42 (21 ref.)

168 CHILDREN--COLONIAL PERIOD

Schmidt, William M. "Health and welfare of Colonial American children." AMER J DIS CHILD 130(7) Jul '76, 694-701 (51 ref.)

169 CHILDREN'S LITERATURE--18th CENTURY

Kiefer, Monica. "Eighteenth century children through their books." AMER J DIS CHILD 130 (7) Jul '76, 726-37 (50 ref.)

170 CHLORPROMAZINE

"The introduction of Chlorpromazine." HOSP COMMUNITY PSYCHIATRY 27(7) Jul '76, 505

171 CHOREA--19th CENTURY

Greenwood, Ronald D. "Chorea: a case report." MED SOC NJ J 73(3) Mar '76, 246-47

172 CITIES AND TOWNS

Friedlander, Gordon D. "In support of mega-

structures." IEEE SPECTRUM 13(7) Jul '76,
36-40

173 -----

Gibson, J. E. "The people 'Doughnut'." IEEE
SPECTRUM 13(7) Jul '76, 50-54

174 -----

Lindgren, Nilo. "Soleri's arcologies." IEEE
SPECTRUM 13(7) Jul '76, 42-45

175 -----

Lindgren, Nilo. "Suburbia: the compact city."
IEEE SPECTRUM 13(7) Jul '76, 46-48

176 -----

Lindgren, Nilo. "What is a city?" IEEE SPEC-
TRUM 13(7) Jul '76, 30-35

177 -----

Vance, James E., Jr. "Cities in the shaping
of the American nation." J GEOGRAPH 75(1)
Jan '76, 41-52

178 -----

Wilson, Peter S. "Dispos-a-city." IEEE SPEC-
TRUM 13(7) Jul '76, 56-63

-----. See also RAPID TRANSIT 763

179 CITIES AND TOWNS--ENERGY REQUIREMENTS
Kaplan, Gadi. "Cities: energy gluttons." IEEE
SPECTRUM 13(7) Jul '76, 72-73

180 CITIES AND TOWNS --INDUSTRIES
Mohl, Raymond A. "The industrial city." EN-
VIRONMENT 18(5) Jun '76, 28-38 (54 ref.)

181 CLARK, ALVAN
"Alvan Clark (1804-1887)." ASTRONOMY 4(7)
Jul '76, 56

CLEFT LIP. See CLEFT PALATE 182

182 CLEFT PALATE
        Rogers, Blair O. "Treatment of cleft lip and
        palate during the Revolutionary War: Bicen-
        tennial reflections." CLEFT PALATE J 13(4)
        Oct '76, 371-90 (54 ref.)
183 CLINICOPATHOLOGICAL CONFERENCES
        Estes, J. Worth. "An 18th century clinico-
        pathologic correlation." NY ACAD MED BULL 52(5)
        Jun '76, 617-26 (13 ref.)
184 -----
        Sanchez, Guillermo. "Case records of the Mas-
        sachusetts General Hospital, weekly clinico-
        pathological exercises, case 51-1976." N
        ENGL J MED 295(25) Dec 16, '76, 1421-29 (22
        ref.)
185 CLOTHES DRYER
        "The clothes dryer: born of winter." EXXON
        CHEM MAG 9(3) '76, 13
186 COAL MINES AND MINING
        Blakely, J. Wes. "America's past and future
        are inextricably tied to coal." COAL MIN
        PROCESS 13(7) Jul '76, 42-51
187 -----
        Mason, Richard H. "An industry thwarted, but
        pushing ahead." COAL MIN PROCESS 13(7) Jul
        '76, 52-56
188 COCHRAN, JOHN
        Rogers, Fred B. "Dear Doctor Bones: John
        Cochran, military surgeon." MED SOC NJ J
        73(3) Mar '76, 237
189 COLDEN, CADWALLADER
        Hoermann, Alfred R. "Cadwallader Colden and
        the mind-body problem." BULL HIST MED 50(3)

Fall '76 392-404 (62 ref.)

COLUMBUS, CHRISTOPHER. See AMERICA--DISCOVERY
AND EXPLORATION 26

190 COMETS--COLONIAL PERIOD
"Comets and transits: 1620-1776." ASTRONOMY
4(7) Jul '76, 23-31

191 COMMUNICABLE DISEASES
Wishnow, Rodney M., and Jesse L. Steinfeld.
"The conquest of the major diseases in the
United States: a Bicentennial retrospect."
ANN REV MICROBIOL 30, '76, 427-50 (57 ref.)

192 COMMUNICATION--PICTORIAL WORKS
"The age of communication." AMER SOC INFO
SCI BULL 2(8) Mar '76, 14-15

193 COMMUNITY MENTAL HEALTH SERVICES
"A century of debate surrounds community
care." HOSP COMMUNITY PSYCHIATRY 27(7) Jul
'76, 490

194 -----
"The community mental health centers." HOSP
COMMUNITY PSYCHIATRY 27(7) Jul '76, 506-7

COMPUTERS. See ENIAC 244

195 COMPUTERS--ANECDOTES, FACETIAE, SATIRE, ETC.
Granholm, Jackson. "Great moments in the
history of computing: December 12, 1777."
DATAMATION 22(7) Jul '76, 71-74

196 COMPUTERS--INFORMATION USE
Salton, Gerard. "Computers and information
science." AMER SOC INFORM SCI BULL 2(8) Mar
'76, 19-21

COMSTOCK LODE. See MINES AND MINERAL RE-
SOURCES--NEVADA 605

197 CONCORD, BATTLE OF, 1775
        "Answering freedom's call." WALLACES' FARM
        101(12) Jun 26, '76, 7

198 CONCORD, MASSACHUSETTS
        Brooks, Paul. "Concord: first town in the
        wilderness." LIV WILDN 40(133) Apr-Jun '76,
        4-9

199 CONESTOGA WAGONS
        Pilley, Mary Male. "Boat-shaped wagon en-
        dures rugged terrain." DESIGN N 31(13) Jul 4,
        '76, 26-27

200 CONFEDERATE STATES MEDICAL AND SURGICAL JOURNAL
        (PERIODICAL)
        Sharpe, William D. "The Confederate States
        Medical and Surgical Journal: 1864-1865."
        NY ACAD MED BULL 52(3) Mar-Apr '76, 373-418
        (127 ref.)

     CONFEDERATE STATES OF AMERICA--NAVY--MEDICAL
     ASPECTS. See SPOTSWOOD, WILLIAM A.W. 837

201 CONSTRUCTION INDUSTRY
        Volpe, S. Peter. "Development of our con-
        struction industry." MIL ENG 68(443) May-Jun
        '76, 202-3

202 COOKERY--COLONIAL PERIOD
        Bennion, Marion. "Food preparation in Colon-
        ial America." AMER DIETET ASSN J 69(1) Jul
        '76, 16-23 (17 ref.)

203 COOPER, THOMAS
        Davenport, Derek A. "Reason and relevance,
        the 1811-13 lectures of Professor Thomas
        Cooper." J CHEM ED 53(7) Jul '76, 419-22 (8
        ref.)

204 COOPERATION--INFORMATION SERVICES
     Werdel, Judith A., and Scott Adams. "U.S.
     participation in world information activi-
     ties." AMER SOC INFORM SCI BULL 2(8) Mar '76,
     44-48
205 CORN
     Stasch, Ann R. "Indian corn: a cheap and nu-
     tritious food-a Bicentennial note." AMER
     DIETET ASSN J 69(2) Aug '76, 137 (1 ref.)
206 -----
     Wennblom, Ralph D. "How corn became our
     greatest crop." FARM J 100(6) May '76, 18-20
207 -----
     Zimmerman, Richard. "Bicentennial corn."
     ORGANIC GARD FARM 23(4) Apr '76, 68-69
208 CORONADO, FRANCISCO VASQUEZ de
     "Coronado and the seven cities of gold."
     NAT PARKS & CON MAG 50(2) Feb '76, 8-10
209 CORONERS AND MEDICAL EXAMINERS--NEW YORK CITY
     Blinderman, Abraham. "The coroner describes
     the manner of dying in New York City, 1784-
     1816." AMER J MED 61(1) Jul '76, 103-10
     (17 ref.)
210 CORPORATIONS
     Novick, Sheldon. "The corporate heart."
     ENVIRONMENT 17(9) Dec '75, 18-20, 25-31
211 COSMOLOGY
     "Expanding universe." ASTRONOMY 4(7) Jul '76,
     90-92
     COSTE, JEAN FRANCOIS. See UNITED STATES--REVO-
     LUTIONARY WAR--FRENCH PARTICIPATION 981
212 COTTON
     Scruggs, C.G. "Cotton, the 'elegant white

rose.'" <u>PROGRESS</u> <u>FARM</u> 91(7) Jul '76, 54-56, 58

COUNT RUMFORD. See THOMPSON, BENJAMIN 902

213 COWS--NUTRITION

McCullough, M.E. "Milestones in dairy cattle feeding." <u>HOARD'S</u> <u>DAIRY</u> 121(13) Jul 10, '76, 793, 812

214 CRITICAL CARE NURSING--FORECASTS

Voorman, Dorothy. "Critical care nursing." <u>RN</u> 39(9) Sep '76, 21-23, 26, 28-29, 32, 34, 38

215 Cushing, Harvey

Davidoff, Leo M. "Living with Harvey Cushing, M.D." <u>NY</u> <u>STATE</u> <u>J</u> <u>MED</u> 76(12) Dec '76, 2214-17

216 DAIRY FARMS--17th CENTURY

Hamel, Ron. "The growth of dairy farming in early America." <u>HOARD'S</u> <u>DAIRY</u> 121(13) Jul 10, '76, 791, 804-5

217 DAIRY INDUSTRY AND TRADE--CHRONOLOGY

"Happy birthday America." <u>DAIRY</u> <u>ICE</u> <u>CREAM</u> <u>FIELD</u> 159(7) Jul '76, 26-29; (8) Aug '76, 44-46, 50-51; (9) Sep '76, 51B, 51D, 52, 54

218 DALLAS, TEXAS

McManus, Michael J. "Back to the people: Dallas' blueprint for the future." <u>CIVIL</u> <u>ENG</u>--<u>ASCE</u> 46(7) Jul '76, 57-60

219 DATA PROCESSING MACHINES--PICTORIAL WORKS

"Automation and computers, the marvelous machines." <u>AMER</u> <u>SOC</u> <u>INFORM</u> <u>SCI</u> <u>BULL</u> 2(8) Mar '76, 22-23

DAVIS, ISAAC. See CONCORD, BATTLE OF, 1775 197

220 DE BRAHM, WILLIAM GERARD

De Vorsey, Louis, Jr. "William Gerard De

Brahm, Surveyor-General and man of science
in Royal Georgia." GA ACAD SCI BULL 34(4)
Sep '76, 204-9 (7 ref.)

221 DE SOTO, HERNANDO
"Hernando De Soto and the Golden Pole." NAT
PARKS & CON MAG 50(1) Jan '76, 13-15

DEATH--CAUSES. See CLINICOPATHOLOGICAL CONFER-
ENCES 183-84; CORONERS AND MEDICAL EXAMIN-
ERS--NEW YORK CITY 209

222 DECLARATION OF INDEPENDENCE
Bishop, Jim. "The week America was born."
MOD MATURITY 19(1) Feb-Mar '76, 31-41

-----. See also FOURTH OF JULY--WEATHER 333;
HALL, LYMAN 378; PHYSICIANS--18th CENTURY
709

223 DENTISTRY--CHRONOLOGY
"American firsts in dentistry." AMER DENT
ASSN J 93(1) Jul '76, 35, 36, 45, 46, 52, 70

224 DERMATOLOGY
Lobitz, Walter C. "Major contributions of
American dermatologists--1926 to 1976." ARCH
DERM 112 (Spec. Issue) Nov 29, '76, 1646-50

225 -----
Shelley, Walter B. "Major contributors to
American dermatology--1876 to 1926." ARCH
DERM 112 (Spec. Issue) Nov 29, '76, 1642-46

226 DERMATOLOGY--NEW YORK
Blau, Saul. "Origins of dermatology in New
York." NY STATE J MED 76(7) Jul '76, 1174-76

227 DIE CASTING
Curry, Gordon C. "Die casting comes to Hoo-
ver." DIE CAST ENG 20(4) Jul '76, 28-32

228 -----

"The custom die casters." DIE CAST ENG 20(4)
Jul '76, 38-42

229 -----

"The golden age of die casting." DIE CAST
ENG 20(4) Jul '76, 24-26

230 -----

"Some pioneers in the U.S. die casting industry." DIE CAST ENG 20(4) Jul '76, 34-37

231 DIET IN DISEASE

Ohlson, Margaret A. "Diettherapy in the U.S.
in the past 200 years." AMER DIETET ASSN J
69(5) Nov '76, 490-97 (73 ref.)

232 DIFFRACTION GRATINGS

Harrison, George R., and Erwin G. Loewen.
"Ruled gratings and wavelength tables." APP
OPTICS 15(7) Jul '76, 1744-47 (22 ref.)

233 DIGITALIS

Burchell, Howard B. "Coincidental Bicenteni-
als: United States and foxglove therapy."
J HIST MED 31(3) Jul '76, 292-306 (59 ref.)

234 -----

Friend, Dale G. "Digitalis after two centuries." ARCH SURG 111(1) Jan '76, 14-19
(28 ref.)

235 DIPHTHERIA

"Cynanche Angina-Connecticut." MORBID MORTAL
WEEK REP 25(25 pt. 2) Jul 2, '76, 7

236 DIX, DOROTHEA LYNDE

"Dorothea Lynde Dix: crusader on behalf of
the mentally ill." HOSP COMMUNITY PSYCHIATRY
27(7) Jul '76, 471-72

237 DRAPER, JOHN WILLIAM
     Hyde, W. Lewis. "John William Draper 1811-
     1882, photographic scientist." APP OPTICS
     15(7) Jul '76, 1726-30 (14 ref.)
238 DRAWINGS--PRE-COLUMBIAN
     Wellman, Klaus F. "Ships on the rocks."
     OCEANS 9(3) May-Jun '76, 8-11
239 DRUG STORES
     Bender, George A. "A Bicentennial salute to
     the American drug store." AMER DRUG 174(1)
     Jul '76, 55-57
     DRUGS. See also THERAPEUTICS 900-01; VETERIN-
     ARY DRUGS 996
     DRUGS--18th CENTURY. See BOTANY, MEDICAL 112;
     MEDICINE--18th CENTURY 566
240 DRUGS--LAWS AND REGULATIONS
     Abrams, William B. "Therapeutics and govern-
     ment: 1776 and 1976." CLIN PHARMACOL THER
     20(1) Jul '76, 1-5 (10 ref.)
241 DUTTON, CHARLES
     Melnick, John C. "Dr. Charles Dutton (1777-
     1842) Youngstown's first doctor." OHIO STATE
     MED J 72(3) Mar '76, 187-88
242 DYNAMICAL SYSTEMS
     Kaplan, James, and Aaron Strauss. "Dynamical
     systems: Birkhoff and Smale." MATH TEACHER
     69(6) Oct '76, 495-501 (7 ref.)
243 DYSENTERY
     "The Bloody Flux-Massachusetts." MORBID MOR-
     TAL WEEK REP 25(25 pt. 2) Jul 2, '76, 4
244 ENIAC
     Strauss, Aaron. "ENIAC: the first computer."
     MATH TEACHER 69(1) Jan '76, 66-72 (14 ref.)

245 EAGLES
     Zimmerman, David R. "The bald eagle Bicenten-
     nial blues." NAT HIST 85(1) Jan '76, 8, 10,
     14, 16
     EARLE, PLINY. See PSYCHIATRY 740
246 EARTHQUAKES
     "Quakes in search of a theory." MOSAIC 7(4)
     Jul-Aug '76, 2-11
247 EDISON, THOMAS A.
     Vanderbilt, Byron M. "America's first R & D
     center." IND RES 18(12) Nov 15, '76, 27-31
     (5 ref.)
248 ELECTRIC CABLES, SUBMARINE
     Finn, Bernard S. "Growing pains at the cross-
     roads of the world: a submarine cable station
     in the 1870's." IEEE PROC 64(9) Sep '76,
     1287-92 (20 ref.)
249 ELECTRIC CURRENTS, ALTERNATING
     Reynolds, Terry S., and Theodore Bernstein.
     "The damnable alternating current." IEEE
     PROC 64(9) Sep '76, 1339-43 (22 ref.)
250 ELECTRIC MOTORS, INDUCTION
     Alger, Philip L., and Robert E. Arnold.
     "The history of induction motors in America."
     IEEE PROC 64(9) Sep '76, 1380-83 (18 ref.)
251 ELECTRIC POWER PLANTS--NIAGARA FALLS
     Belfield, Robert. "The Niagara system: the
     evolution of an electric power complex at
     Niagara Falls, 1883-1896." IEEE PROC 64(9)
     Sep '76, 1344-50 (29 ref.)
252 ELECTRIC RAILROADS
     Condit, Carl W. "Railroad electrification in
     the United States." IEEE PROC 64(9) Sep '76,

1350-60 (47 ref.)

253 ELECTRIC UTILITIES

Hughes, Thomas P. "Technology and public
policy: the failure of giant power." IEEE
PROC 64(9) Sep '76, 1361-71 (38 ref.)

254 -----

Novick, Sheldon. "The electric power industry." ENVIRONMENT 17(8) Nov '75, 7-13, 32-39

255 ELECTRIC WELDING, ARC

Nunes, A.C., Jr. "Arc welding origins."
WELDING J 55(7) Jul '76, 566-72 (29 ref.)

256 ELECTRICAL ENGINEERING EDUCATION

Terman, Frederick E. "A brief history of
electrical engineering education." IEEE PROC
64(9) Sep '76, 1399-1407 (6 ref.)

257 ELECTRICITY

The foundation years (1754-1837), understanding that nature obeys rules, too." ELECTRON-
IC DESIGN 24(4) Feb 16, '76, 66-74

258 ELECTRICITY--19th CENTURY

"The era of giants (1837-1879), getting
electricity to work for man." ELECTRONIC DE-
SIGN 24(4) Feb 16, '76, 78-84

259 ELECTRONIC DATA PROCESSING--CONSUMER APPLI-
CATIONS

Allan, Roger. "The citizen-city interface."
IEEE SPECTRUM 13(7) Jul '76, 64-71

ELECTRONIC OVENS, BAKING. See STOVES 846

260 ELECTRONICS

Susskind, Charles. "American contributions
to electronics: coming of age and some more."
IEEE PROC 64(9) Sep '76, 1300-05 (25 ref.)

261 ELECTRONICS INDUSTRY
        Norberg, Arthur L. "The origins of the elec-
        tronics industry on the Pacific Coast." IEEE
        PROC 64(9) Sep '76, 1314-22 (28 ref.)
262 ELLICOTT, ANDREW
        Bedini, Silvio A. "Andrew Ellicott, surveyor
        of the wilderness." SURVEY MAP 36(2) Jun '76,
        113-35 (57 ref.)
263 ELMER, JONATHAN
        Rogers, Fred B. "Jonathan Elmer: medical pro-
        genitor." MED SOC NJ J 73(4) Apr '76, 330A
264 EMERGENCY MEDICINE
        Rockwood, Charles A., Jr., Colleen M. Mann,
        J.D. Farrington, Oscar P. Hampton, and Robert
        E. Motley. "History of emergency medical ser-
        vices in the United States." J TRAUMA 16(4)
        Apr '76, 299-308 (45 ref.)
265 EMERGENCY MEDICINE--NEW YORK
        Cameron, Cyril T.M., Edward W. Gilmore, and
        Edward L. McNeil. "History of emergency
        medicine in New York State." NY STATE J MED
        76(7) Jul '76, 1176-78 (4 ref.)
266 EMIGRATION AND IMMIGRATION
        "The golden door." NAT PARKS & CON MAG 50
        (12) Dec '76, 8-10
267 -----
        "A nation of immigrants." METROPOL LIFE STAT
        BULL 57 Nov '76, 2-5
268 EMIGRATION AND IMMIGRATION--HEALTH
        Friedman, Robert. "Stephen Smith and health
        of immigrants, 1850 to 1865." NJ STATE J MED
        76(11) Nov '76, 2050-52 (17 ref.)

269 EMPIRE STATE BUILDING
     Dallaire, Gene. "Empire State: greatest of
     all skyscrapers." CIVIL ENG-ASCE 46(7) Jul
     '76, 65-68
270 ENAMEL AND ENAMELING
     "From art to appliances." CER IND 107(1) Jul
     '76, 26-27, 44
271 ENDODONTICS
     Grossman, Louis I. "Endodontics 1776-1976: a
     Bicentennial history against the background
     of general dentistry." AMER DENT ASSN J 93(1)
     Jul '76, 78-87 (14 ref.)
272 ENERGY--CHRONOLOGY
     Small, Margaret G. "Energy: the first two
     hundred years." PIPELINE & GAS J 203(8) Jul
     '76, 29-30, 32, 34, 36-37, 40, 42, 46, 50,
     52, 54, 56
273 -----
     "200 years of energy." CHILTON'S OIL GAS
     ENERGY 2(1) Jan '76, 22-35
274 ENERGY--FORECASTS
     "Century III report." PIPELINE & GAS J 203(8)
     Jul '76, 18-21, 24-25, 28
275 ENGINEERING
     Bentley, Donald C. "Two centuries of engin-
     eering accomplishments." MIL ENG 68(443)
     May-Jun '76, 204-6
276 -----
     Hirschfeld, Fritz. "Engineering in the new
     nation." MECH ENG 98(5) May '76, 18-19
277 -----
     Parthum, Charles A. "Two hundred years-and-
     more-of engineering in the United States."

ENG ISSUES (PROC ASCE) 102(EI4) Oct '76,
423-28

278   ENGINEERING--FORECASTS
Merdinger, Charles J. "An engineering brief
for the 1980's." MIL ENG 68(443) May-Jun '76,
212-14

279   -----
Zetlin, Lev. "Engineering requirements for
the year 2000." MIL ENG 68(443) May-Jun '76,
215-17

280   ENGINEERING ETHICS
Kiser, James P. "Society's changing perception of the professional and ethical conduct
of the civil engineer: 1776-1976." ENG ISSUE
(PROC ASCE) 103(EI1) Jan '77, 37-40

281   -----
Koepp, Glenn R. "The impact of engineering
ethics on the development of modern civil
engineering practice, 1776-1976." ENG ISSUES
(PROC ASCE) 103(EI1) Jan '77, 1-4 (5 ref.)

282   ENGINEERING SOCIETIES
Rubinstein, Ellis. "IEEE and the Founding
Societies." IEEE SPECTRUM 13(5) May '76,
76-84

283   ENGINEERS
"The engineering profession in Bicentennial
1976." PROF ENG 46(1) Jan '76, 29-44

284   ENGLISH, THOMAS DUNN
Rogers, Fred B. "Thomas Dunn English: doctor,
lawyer, author." MED SOC NJ J 73(5) May '76,
455-56

285   ETOMOLOGICAL SOCIETY OF WASHINGTON
Gurney, Ashley B. "A short history of the

Entomological Society of Washington." <u>ENTOMOL</u>
<u>SOC</u> <u>WASH</u> <u>PROC</u> 78(3) Jul '76, 225-39 (17 ref.)
286 ENTOMOLOGY
Musgrave, C.A., and D.R. Bennett. "Bicentennial review of early American entomology."
<u>FLOR</u> <u>ENTOMOL</u> 59(4) Dec '76, 329-33 (7 ref.)
287 ENVIORNMENTAL POLICY
Strohm, John. "Our times, too, call for great-
ness." NAT WILDLIFE 14(1) Dec-Jan '76, 4-15
288 ERIE CANAL
Langbein, W.B. "Hydrology and environmental
aspects of Erie Canal (1817-99)." <u>US</u> <u>GEOL</u>
<u>SURV</u> <u>WATER</u> <u>SUPPLY</u> <u>PAP</u> (2038) '76, (67 ref.)
ESPY, JAMES P. See WEATHER 1033
289 FAMILIES--FOOD AND NUTRITION
Lowenberg, Miriam E., and Betty L. Lucas.
"Feeding families and children--1776 to 1976."
<u>AMER</u> <u>DIETET</u> <u>ASSN</u> <u>J</u> 68(3) Mar '76, 207-15
(37 ref.)
290 FARM LIFE--18th CENTURY
Westbrook, Perry D. "Farm life in Colonial
New York." CONSERVATIONIST 30(5) Mar-Apr '76,
10-13
291 FARM PRODUCE--MARKETING
Tontz, Robert L. "200 years of U.S. farm
trade policy." <u>FOR</u> <u>AGRI</u> 14(42) Oct 18 '76,
6-8, 10, 14; (43) Oct 25 '76, 6-8, 12
-----. See also AGRICULTURE 10
292 FARMERS
Black, Roe C. "The American farmer . . . .
our first 'hybrid'." <u>FARM</u> <u>J</u> 100(2) Feb '76,
24-26, 64
-----. See also Washington, George 1013

293 FARMING--18th CENTURY
     "Step back to '76." NAT FUTURE FARM 24(5)
     Jun-Jul '76, 28-29
294 FARMING--19th CENTURY
     Brinkman, Grover. "Footsteps to the past."
     AMER AGRICULTURIST 173(3) Mar '76, 41
295 FARMS
     Machan, Cathy Sherman. "Going on 200
     years . . . . these farms stayed in the
     family." FARM J 100(7) Jun-Jul '76, 23-25
296 FARMS--NEW ENGLAND
     "A visit to some of New England's Bicenten-
     nial farms." HOARD'S DAIRY 121(13) Jul 10,
     '76, 796-97, 800
     FERTILIZERS AND MANURES. See SEWAGE,UTILIZA-
     TION OF 813
297 FEVER
     Smith, Dale C. "Quinine and fever: the de-
     velopment of the effective dosage." J HIST
     MED 31(3) Jul '76, 343-67 (96 ref.)
298 FIRE PROTECTION
     Burns, Robert. "When the watchman spun his
     rattle, the cry was 'throw out your buc-
     kets'." FIRE ENG 129(7) Jul '76, 20-27
     FIREARMS. See UNITED STATES CONTINENTAL ARMY
     953
299 FIREARMS--18th CENTURY
     Kern, Steve. "Americans achieve arms."
     DESIGN N 31(13) Jul 4 '76, 34
300 FIREARMS--PICTORIAL WORKS--18th CENTURY
     "Guns of the Bicentennial." AMER RIFLEMAN
     124(2) Feb '76, 22-25

301 FIREARMS INDUSTRY AND TRADE--18th CENTURY
     Serven, James E. "Gunmaking in Colonial
     America." GUNS 22(6-1) Jan '76, 46-47, 72,
     73
302 FIRES
     Lyons, Paul R. "Our heritage of fire in
     America." FIRE J 70(4) Jul '76, 41-45, 53-54
303 FISHES
     "Exotic fishes in United States waters."
     TENN CONSERV 42(6) Jul '76, 5-7
304 FITZSIMMONS, JAMES E., "SUNNY JIM"
     Fitzsimmons, John J. "The best years of our
     lives." HORSEMEN'S J 27(2) Feb '76, 24-28
305 FLAGS
     Clepper, Henry. "America's tree flags." AMER
     FOR 82(3) Mar '76, 22-25, 67
306 FLOUR MILLS
     "George Washington's grist mill." SCI & CHILD
     13(4) Jan '76, 29-30
307 -----
     Stefanides, E. J. "Western wilderness gets
     first modern grist mill." DESIGN N 31(13)
     Jul 4, '76, 28-29
     FLU. See INFLUENZA 431
308 FOOD
     Enloe, Cortez F., Jr. "De ship da hab de
     bush." NUTR TODAY 10(3) May-Jun '75, 5-15
     (9 ref.)
     FOOD--LAW AND REGULATION. See FOOD ADULTERA-
     TION AND INSPECTION 311
309 FOOD--PICTORIAL WORKS
     Enloe, Cortez F., Jr. "From the eye inward."
     NUTR TODAY 11(1) Jan-Feb '76, 20-24

310   FOOD-PACKAGING--18th CENTURY
         Barol, Leonard L. "Some observations on food
         packaging and distribution in the Revolution-
         ary period." FOOD TECH 30(2) Feb '76, 54, 56
311   FOOD ADULTERATION AND INSPECTION
         Day, Harry G. "Food safety--then and now."
         AMER DIETET ASSN J 69(3) Sep '76, 229-34 (13
         ref.)
312   FOOD INDUSTRY
         Hall, Ross Hume. "The food fabricators." EN-
         VIRONMENT 18(1) Jan-Feb '76, 25-34 (33 ref.);
         (2) Mar '76, 17-20, 25, 32-36 (18 ref.)
313   -----
         "The twentieth century--the industry ma-
         tures." FOOD SERV MKTG 38(7) Jul '76, 64-65,
         68, 71-72, 74, 76, 78-79
314   FOOD INDUSTRY--19th CENTURY
         "A nostalgic look at the nineteenth century."
         FOOD SERV MKTG 38(7) Jul '76, 39-40, 44-45,
         48, 52, 54, 56-57, 60
315   FOOD INDUSTRY--LAWS AND REGULATIONS
         Middlekauff, Roger D. "200 years of U.S. food
         laws: a Gordian knot." FOOD TECH 30(6) Jun
         '76, 48, 50, 52, 54 (21 ref.)
316   FOOD MARKETING--18th CENTURY
         McClelland, Charles W. "Food marketing in the
         Revolutionary period." FOOD TECH 30(11) Nov
         '76, 82, 84
317   FOOD PRESERVATION--DRYING
         Labuza, Theodore P. "Drying food: technology
         improves on the sun." FOOD TECH 30(6) Jun '76
         37, 38, 42, 44, 46 (23 ref.)

318 FOOD PRESERVATION--FREEZING
Fennema, Owen. "The U.S. frozen food industry: 1776-1976." FOOD TECH 30(6) Jun '76, 56
58-61, 68 (29 ref.)

319 FOOD PRESERVATION--HEATING
Goldblith, Samuel A. "Thermal processing in
retrospect and prospect." FOOD TECH 30(6)
Jun '76, 32-33, 46 (3 ref.)

320 FOOD SERVICE INDUSTRY--BIOGRAPHY
"Bicentennial profiles, 10 pioneers who
shaped modern American foodservice." INST/
VOL FEEDING 79(1) Jul 1, '76, 27-33, 41-42,
44

321 FOOD SUPPLY--18th CENTURY
Todhunter, E. Neige. "Nutrition and the food
supply during the Revolutionary period."
FOOD TECH 30(7) Jul '76, 32, 34
FORENSIC PSYCHIATRY. See RAY, ISAAC 764

322 FOREST MANAGEMENT
Hatfield, Mark O. "The present resource sit-
uation: 'waiting for a crisis'." AMER FOR
82(2) Feb '76, 8-9, 68-69

323 -----
Hitch, Charles J. "Resources for 300 mil-
lion." AMER FOR 82(4) Apr '76, 8-9, 60-63

324 FORESTS AND FORESTRY
Butz, Earl L. "Our nation's forests." AMER
FOR 82(1) Jan '76, 10-11, 54

325 FORESTS AND FORESTRY--PICTORIAL WORKS
"What forests mean to America." AMER FOR
82(1) Jan '76, 24-25; (2) Feb '76,46-47; (3)
Mar '76, 44-45; (4) Apr '76, 34-35; (5) May
'76, 40-41; (6) Jun '76, 40-41; (7) Jul '76

40-41; (8) Aug '76, 32-33; (9) Sep '76, 40-
41; (10) Oct '76, 44-45; (11) Nov '76, 42-43;
(12) Dec '76, 40-41

326   FORT, GEORGE FRANKLIN
      Rogers, Fred B. "George Franklin Fort: New
      Jersey's first physician-governor." MED SOC
      NJ J 73(9) Sep '76, 747

327   FORT SNELLING (MINNESOTA) HOSPITAL
      Wiggins, David S. "Minnesota's first hos-
      pital, the practice of medicine at Ft. Snel-
      ling 1819-1840." MINN MED 59(12) Dec '76,
      867-73, 886 (14 ref.)

328   FORT TICONDEROGA
      Billard, Tory. "The guns of Ticonderoga."
      TRANSLOG 7(7) Jul '76, 1-5

329   FOUNDRIES
      Gibson, Susan L. "Castings and the molding
      of America." FOUNDRY MGT & TECH 104(1) Jan
      '76, 34-36, 38, 40; (2) Feb '76, 78, 80, 82;
      (3) Mar '76, 110-11, 113, 114; (4) Apr '76,
      203-4, 206, 208; (5) May '76, 108-10, 112,
      114-15; (6) Jun '76, 164-66, 168, 170; (7)
      Jul '76, 84-85, 88

330   -----
      "History of metalcasting in America." MOD
      CAST 66(1) Jan '76, 42; (2) Feb '76, 53;
      (3) Mar '76, 63; (4) Apr '76, 60; (5) May
      '76, 63; (6) Jun '76, 63; (7) Jul '76, 42;
      (8) Aug '76, 39; (10) Oct '76, 38; (11)
      Nov '76, 87

331   FOUNDRIES--ANECDOTES, FACETIAE, SATIRE, ETC.
      "Face to face--Benjamin Franklin interviews
      Paul Revere." MOD CAST 66(7) Jul '76, 30-31

332   FOUNDRIES--COMPANIES
         "Cavalcade of craftmanship-honoring perse-
         verance." <u>MOD</u> <u>CAST</u> 66(7) Jul '76, 32-41; (8)
         Aug '76, 46-49; (12) Dec '76, 70-71
333   FOURTH OF JULY--WEATHER
         "Independence day weather-1776." <u>WEATHERWISE</u>
         28(3) Jun '75, 107
      FOXGLOVE. See DIGITALIS 233-34
334   FRANKLIN, BENJAMIN
         Andreasen, N. J. C. "Benjamin Franklin:
         physicus et medicus." <u>AMER</u> <u>MED</u> <u>ASSN</u> <u>J</u> 236
         (1) Jul 5, '76, 57-62
335   -----
         "Benjamin Franklin." <u>DESIGN</u> <u>N</u> 31(13) Jul 4,
         '76, 13-15
336   -----
         Finn, Bernard S., "Franklin as electrician."
         <u>IEEE</u> <u>PROC</u> 64(9) Sep '76, 1270-73 (16 ref.)
337   -----
         Greenblatt, Robert B. "The autumnal years of
         Benjamin Franklin." <u>GERIATRICS</u> 31(7) Jul '76,
         100-102
338   -----
         Heilbron, John L. "Franklin's physics."
         <u>PHYS</u> <u>TODAY</u> 29(7) Jul '76, 32-37
339   -----
         James, John W. "Benjamin Franklin's contri-
         butions to the art of heating and ventila-
         ting." <u>ASHRAE</u> <u>J</u> 18(7) Jul '76, 47-48
340   -----
         McMahon, A. Michal. "Benjamin Franklin:
         revolutionary and experimentalist." <u>SCI</u> <u>&</u>
         <u>CHILD</u> 13(4) Jan '76, 33-34

341 -----
     Moyer, Albert E. "Benjamin Franklin: 'let
     the experiment be made'." PHYS TEACH 14(9)
     Dec '76, 536-45 (9 ref.)
342 -----
     Pomerantz, Martin A. "Benjamin Franklin--the
     complete geophysicist." AMER GEOPHYS UN
     TRANS 57(7) Jul '76, 492-505
343 -----
     Sherr, Virginia T. "Benjamin Franklin and
     geropsychiatry: vignettes for the Bicenten-
     nial year." AMER GERIAT SOC J 24(10) Oct '76,
     447-51 (14 ref.)
     -----. See also MEDICINE--COLONIAL PERIOD 560;
     PHYSICAL MEDICINE 703; RUSH, BENJAMIN 779
344 FRANKLIN, BENJAMIN--ANECDOTES, FACETIAE, SA-
     TIRE, ETC.
     Park, Edward. "'Absolutely, Dr. Franklin?'
     'Positively Mr. Jefferson!'" SMITHSONIAN
     7(4) Jul '76, 50-51
345 FRANKLIN INSTITUTE. JOURNAL (PERIODICAL)
     Pomerantz, Martin A. "Journal of the Frank-
     lin Institute: 'for the diffusion of scientific
     of knowledge'." FRANKLIN INST J 301(1-2)
     Jan-Feb '76, 7-25
346 FRONTIER AND PIONEER LIFE--MEDICAL ASPECTS
     Brieger, Gert H. "Health and disease on the
     western frontier." WEST J MED 125(1) Jul '76,
     28-35
     FRONTIER AND PIONEER LIFE--TENNESSEE. See
     SYCAMORE SHOALS, TENNESSEE--18th CENTURY
     859

347 FRUIT INDUSTRY AND TRADE
     "A Bicentennial celebration of fresh fruit
     and vegetables." UNITED FRESH FRUIT & VEGE-
     TABLE ASSN ANNUAL '76, 71, 74, 76, 78, 81,
     82, 84, 303, 304, 306-13
348 FRUIT INDUSTRY AND TRADE--PICTORIAL WORKS
     "A Bicentennial tribute to yesterday's fruit
     growers." AMER FRUIT GROW 96(7) Jul '76, 10
349 FULTON, FRANK TAYLOR
     Gilman, John F. W. "Frank Taylor Fulton,
     M.D., 1867-1961." RHODE ISLAND MED J 59(3)
     Mar '76, 113-17, 138-40
350 GARCES, FRANCISCO
     Mizell, Mary. "The spirit of Arizona--1776."
     ARIZONA HIGHWAYS 52(1) Jan '76, 11-15
351 GARDENING
     Kraft, Ken and Pat. "'Dear gardener,' the ad-
     vice of long ago." FLOWER GARD 20(1) Jan '76,
     26-27, 54
     -----. See also JEFFERSON, THOMAS 462, 469
352 GARDENS--COLONIAL PERIOD
     Macneale, Peggy. "The flowers of Colonial
     gardens." FLOWER GARD 20(5) May '76, 22-25
353 -----
     Wilson, Helen van Pelt. "18th century flow-
     ers." HORTICULTURE 54(6) Jun '76, 32, 34, 36
354 GARDENS--COLONIAL PERIOD--FENCES
     Burgess, Lorraine. "A study of Colonial gar-
     den fences." FLOWER GARD 20(6) Jun '76,
     24-25
355 GARFIELD, JAMES
     Graham, Malcolm. "President Garfield and the
     Pythagorean theorem." MATH TEACHER 69(8)

Dec '76, 686-87 (5 ref.)

356 GAS, NATURAL--INDUSTRY

"The story of gas." AMER GAS ASSN MO 57(7-8)
Jul-Aug '75, 22-23; (9) Sep '75, 17-18; (10)
Oct '75, 14-15; (11) Nov '75, 26-27; (12)
Dec '75, 40-41; and 58(1) Jan '76, 12-13; (2)
Feb '76, 14-15; (3) Mar '76, 12-13; (4) Apr
'76, 26-27; (5) May '76, 16-17; (6) Jun '76,
26-27; (7-8) Jul-Aug '76, 24-25

357 GENERAL PRACTICE--NEW YORK

Berger, Herbert. "History of Academy of
Family Physicians in New York State." NY
STATE J MED 76(7) Jul '76, 1182-84

358 GEODESY

Berry, Ralph Moore. "History of geodetic
leveling in the United States." SURVEY MAP
36(2) Jun '76, 137-53 (68 ref.)

-----. See also SURVEYING--18th CENTURY 856

359 GEOGRAPHY, CULTURAL

Holtgrieve, Donald G. "Land speculation and
other processes in American historical geography." J GEOGRAPH 75(1) Jan '76, 53-64
(38 ref.)

360 GEOLOGY

Hazen, Robert M. "The founding of geology in
America: 1771 to 1818." GEOL SOC AMER BULL
85(12) Dec '74, 1827-33 (46 ref.)

361 GEOLOGY--BOSTON--COLONIAL PERIOD

Kaye, Clifford A. "The geology and early history of the Boston area of Massachusetts, a
Bicentennial approach." US GEOL SURV BULL
(1476) '76, 1-78 (29 ref.)

GEOPHYSICS. See FRANKLIN, BENJAMIN 342

362 GEORGE III, KING OF ENGLAND
        Taylor, Blaine. "England's George III: The
        Mad King--a medical casefile." MD STATE MED
        J 25(7) Jul '76, 35-41 (29 ref.)
363 GERIATRIC NURSING--FORECASTS
        Stone, Virginia. "The nurse and the aged."
        RN 39(2) Feb '76, 21-22, 24, 26
     GERIATRICS. See FRANKLIN, BENJAMIN 337, 343
364 GETTYSBURG, BATTLE OF, 1863
        Whitehouse, Walter M., and Frank Whitehouse,
        Jr. "The daily register of Dr. Cyrus Bacon,
        Jr.: care of the wounded at the battle of
        Gettysburg." MICH ACAD 8(4) Spr '76, 373-86
        (17 ref.)
365 GIBBS, JOSIAH WILLARD
        Morowitz, Harold J. "Let free energy ring."
        HOSP PRACT 11(1) Jan '76, 189-90
        -----. See also PHASE RULE AND EQUILIBRIUM 695
366 GLASS
        Martin, David M. "Revolutions in glassmak-
        ing." GLASS IND 57(6) Jun '76, 28, 30, 33,
        34 (3 ref.)
367 -----
        Simpson, H.E. "Glass--our early starter."
        CER IND 107(1) Jul '76, 22-23
        -----. See also AMERICAN CERAMICS SOCIETY--
        GLASS DIVISION 27
     GOLD MINES AND MINING. See MINES AND MINERAL
        RESOURCES--NEVADA 605
368 GORGAS, WILLIAM CRAWFORD
        Breunle, Phillip C. "William Crawford Gorgas:
        military sanitarian of the Isthmian Canal."
        MIL MED 141(11) Nov '76, 795-97 (12 ref.)

369 GRAIN--ANECDOTES, FACETIAE, SATIRE, ETC.
       Mattson, Jeffrey. "The history of cereal in
       America." CEREAL FOOD WORLD 21(7) Jul '76,
       320
     GRAY, ELISHA. See BELL, ALEXANDER GRAHAM 100
370 GREENE, NATHANAEL
       Grenier, Fred. "The private's strategy."
       SOLDIERS 31(5) May '76, 42-43
     GRETH, AUGUST. See AIRSHIPS 19
371 GRIDLEY, RICHARD
       Buzzaird, Raleigh B. "America's first chief
       engineer." MIL ENG 68(441) Jan-Feb '76, 31-
       36
     GRIST MILLS. See FLOUR MILLS 306-7
372 GROUP MEDICAL PRACTICE
       Custer, G. Stanley. "The development of group
       practice in the Midwest and Wisconsin." WIS
       MED J 75(7) Jul '76, 32-37 (20 ref.)
     GUNPOWDER. See UNITED STATES--CONTINENTAL ARMY
       953
373 GUNTER, EDMUND
       Steward, Harry. "Gunter, Gunther, and Gün-
       ther: some notes in the history of survey-
       ing." AMER CONG SURVEY MAP BULL (52) Feb '76,
       7-9 (24 ref.)
     GYNECOLOGY--NEW YORK. See OBSTETRICS 667
     GYNECOLOGY--RHODE ISLAND--19th CENTURY. See
       PORTER, GEORGE WHIPPLE 729
374 HALE, GEORGE ELLERY
       "George Ellery Hale (1868-1938)." ASTRONOMY
       4(7) Jul '76, 93
375 HALL, CHARLES MARTIN
       Russell, Allen S. "Charles Martin Hall, first

chemist in the Inventors Hall of Fame."
CHEMTECH 6(2) Feb '76, 83-85

376 HALL, EDWIN

Moyer, Albert E. "Edwin Hall and the emer-
gence of the laboratroy in teaching physics."
PHYS TEACH 14(2) Feb '76, 96-103 (14 ref.)

377 HALL, JAMES

Fisher, Donald W. "James Hall." CONSERVATION-
IST 31(31) Nov-Dec '76, 12-16

378 HALL, LYMAN

Williams, W. Talbert. "Dr. Lyman Hall, signer
of the Declaration of Independence." MED ASSN
GEORGIA J 65(7) Jul '76, 277-79 (5 ref.)

HAMMOND, WILLIAM ALEXANDER. See NEUROLOGY 642

379 HANDICRAFTS--EXHIBITIONS

"Bicentennial bagatelle." CRAFT HORIZONS 36
(3) Jun '76, 32-36, 53-59

380 HARDWARE

"The great American hardware story." HARDWARE
AGE 212(7) Jul 4, '75, 8-12, 14, 16, 18, 20,
24, 33, 35, 39-40, 43-44, 48, 50, 55-56, 58,
60, 62-63, 66, 68-70, 72-80, 86, 88, 92, 95-
96, 98, 100-102, 104, 106, 108, 110, 112, 114

381 HARDWARE--PICTORIAL WORKS

"Yesterday's new products." HARDWARE AGE 212
(7) Jul 4, '75, 124-41

382 HARDWARE AGE (PERIODICAL)

"120 years of Hardware Age history." HARDWARE
AGE 212(7) Jul 4, '75, 118-22

HARRIS, GRANDISON, SR. See HUMAN DISSECTION 409

383 HARVESTING MACHINERY

"Harvest machines give you the power of 1,000
men." FARM J 100(5) Apr '76, F-8

384 HASSLER, RUDOLPH
        Stanley, Albert A. "Hassler's legacy." NOAA
        6(1) Jan '76, 52-57
385 HEART DISEASES
        Paul, Oglesby. "Cardiac disease." ANN INTERN
        MED 85(4) Oct '76, 520-21
        HEART DISEASES. See also CARDIOLOGY 133-34
386 HENRY, JOSEPH
        Molella, Arthur P. "The electric motor, the
        telegraph, and Joseph Henry's theory of tech-
        nological progress." IEEE PROC 64(9) Sep '76,
        1273-78 (40 ref.)
387 HERTZLER, ARTHUR
        Oppenheim, Elliott B. "Surgery beneath the
        Kansas apple tree: Arthur Hertzler, M.D."
        REV SURG 33(3) May-Jun '76, 149-51 (3 ref.)
388 HESSIAN MERCENARIES--REVOLUTIONARY WAR--MEDICAL
        ASPECTS
        Schmitz, Rudolf. "Hessian troops in the Amer-
        ican Revolution, their medical and medicinal
        care." MINN MED 59(7) Jul '76, 479-82
389 HILL, LUTHER LEONIDAS, JR.
        "Luther Leonidas Hill, Jr., M.D." MED ASSN
        STATE ALA J 45(12) Jun '76, 22-24 (5 ref.)
390 HOMEOPATHY
        Flinn, Lewis B. "Homeopathic influence in the
        Delaware community: a retrospective reassess-
        ment." DEL MED J 48(7) Jul '76, 418-25, 427-
        28 (28 ref.)
391 HORSE RACING
        Moore, Robert. "First things first." HORSE-
        MEN'S J 27(2) Feb '76, 49

392   HORSE RACING--LAWS AND REGULATIONS
      Pelton, Robert W. "We don't write 'em, mis-
      ter, we just enforce 'em." HORSEMEN'S J 27(2)
      Feb '76, 54-55
393   HORSES
      Denhardt, Bob. "Bicentennial horses." CATTLE-
      MEN 63(1) Jun '76, 58, 88, 90, 92
394   HORSES--VETERINARY MEDICINE
      Kester, Wayne O. "Development of equine vet-
      erinary medicine in the United States." AMER
      VET MED ASSN J 169(1) Jul 1, '76, 50-55
      HORSESHOEING. See UNITED STATES--CONTINENTAL
      ARMY--VETERINARY MEDICINE 966
395   HOSPITAL PHARMACY SERVICE
      Berman, Alex. "American hospital pharmacy:
      a Bicentennial perspective." AMER J HOSP FARM
      33(2) Feb '76, 129-33 (38 ref.)
396   HOSPITALS
      Ellis, Ce. "Hospitals: 200 years of improve-
      ment." MOD HEALTHCARE 6(1) Jul '76, 16m-16o
397   -----
      O'Connor, Robin, "American hospitals: the
      first 200 years." HOSPITALS 50(1) Jan 1, '76,
      62-72 (3 ref.)
398   -----
      Williams, William H. "Independence and early
      American hospitals, 1751-1812." AMER MED ASSN
      J 236(1) Jul 5, '76, 35-39
      -----. See also FORT SNELLING (MINNESOTA) HOS-
      PITAL 327; LENOX HILL HOSPITAL (NEW YORK
      CITY) 482; MANHATTAN EYE, EAR & THROAT HOS-
      PITAL (NEW YORK CITY) 526; MEDICAL CENTERS--
      NEW YORK 546; NEW YORK MEDICAL COLLEGE 646;

PHARMACISTS 691; PSYCHIATRIC HOSPITALS 737-
38; ST. JOHN'S HOSPITAL (SPRINGFIELD, ILLI-
NOIS) 789; ST. VINCENT'S HOSPITAL (NEW YORK
CITY) 790

399 HOSPITALS--MILITARY
Shapiro, Herman H. "The 44th General Hospit-
al." WIS MED J 75(11) Nov '76, 31-33
HOSPITALS--MILITARY--18th CENTURY. See TILTON,
JAMES 903

400 HOSPITALS--MILITARY--NEW JERSEY
Rogers, Fred B. "The old barracks at Trenton:
military hospital of the Revolution." MED
SOC NJ J 73(1) Jan '76, 11-13 (6 ref.)

401 HOSPITALS--NEW YORK
"Hospital patients--New York, 1777." MORBID
MORTAL WEEK REP 25 (25 pt. 2) Jul 2, '76, 4

402 -----
"Hospitals." NY STATE J MED 76(7) Jul '76,
1240-51

403 HOTELS, TAVERNS, ETC.
Finn, Myrle. "Some fine old inns." HOSPITAL-
ITY 60(6) Jun '76, R67-R69, R72, R74
-----. See also RESTAURANTS 768

404 HOUSEHOLD APPLIANCES
"Portable appliances, 500,000 years in the
making." EXXON CHEM MAG 9(3) '76, 10-11

405 HOUSES--MATERIALS
Williams, Franklin E. "Building materials
for residential construction in Colonial
America." CONSTR REV 22(6) Jul '76, 4-17
(174 ref.)

406 HOWE, SAMUEL GRIDLEY
"Samuel Gridley Howe and the education of the

retarded." HOSP COMMUNITY PSYCHIATRY 27(7)
Jul '76, 478-79

407 HUBBLE, EDWIN P.
"Edwin P. Hubble (1889-1953)." ASTRONOMY 4(7)
Jul '76, 94

408 HUBER, FRANCIS
Morse, Grant D. "How about the research work
of Francis Huber?" GLEANINGS 104(1) Jan '76,
8-9

HUDSON RIVER. See OBSTACLES (MILITARY SCIENCE)
664-65

409 HUMAN DISSECTION
Allen, Lane. "Grandison Harris, Sr.; slave,
resurrectionist and judge." GA ACAD SCI BULL
34(4) Sep '76, 192-99

410 HUMAN ECOLOGY
Graf, William L. "Resources, the environment,
and the American experience." J GEOGRAPH 75
(1) Jan '76, 28-40 (43 ref.)

411 HUNTING
Samuel, David E. "Antihunting: 200-year per-
spective in Critical conservation choices: a
Bicentennial look." Proc. 31st Ann Meet Soil
Conserv Soc Amer, Aug '76, Minneapolis, SCSA,
Ankeny, IA. '76, 44-47 (18 ref.)

HUTCHINSON, JAMES. See WISTAR, CASPAR 1052

412 HYDRANTS--BICENTENNIAL CELEBRATIONS, ETC.
"Hydrant art plugs the Bicentennial." PUB
WORKS 107(7) Jul '76, 34-35

413 HYMNS--BICENTENNIAL
Pryor, Hubert C. "Bicentennial birthday
hymn." MOD MATURITY 19(3) Jun-Jul '76, 25

414 HYPNOSIS--MEDICINE
     Quen, Jacques M. "Mesmerism, medicine and
     professional prejudice." NY STATE J MED 76
     (12) Dec '76, 2218-22 (33 ref.)
415 HYSTERIA
     Woolsey, Robert M. "Hysteria: 1875 to 1975."
     DIS NERV SYST 37(7) Jul '76, 379-86 (57 ref.)
416 ICE INDUSTRY--19th CENTURY
     Barnett, S. E. "The American ice harvests."
     ASHRAE J 18(7) Jul '76, 34-35
     IMMIGRATION. See EMIGRATION AND IMMIGRATION
     266-68
417 IMMUNITY, CUTANEOUS
     Baer, Rudolf L. "Cutaneous immunology--then
     and now, reflections on the period 1876 to
     1976." ARCH DERM 112(Spec. Issue) Nov 29, '76,
     1661-65 (64 ref.)
418 INDIAN CHILDREN
     Sayre, James W., and Robert F. Sayre. "Ameri-
     can children and the children of nature."
     AMER J DIS CHILD 130(7) Jul '76, 716-23 (38
     ref.)
419 INDIAN PONIES
     O'Brien, Mary A. "Indian acquisition of the
     horse." WEST HORSE 41(9) Sep '76, 64-65, 126,
     128-30; (10) Oct '76, 38-40
     INDIANS. See IROQUOIS 454
     INDIANS--AGRICULTURE. See AGRICULTURE--PRE-
     COLUMBIAN 15
420 INDIANS OF NORTH AMERICA--ART--PICTORIAL WORKS
     Guy, Hubert. "Traditional use of patriotic
     designs in American Indian art." ARIZONA
     HIGHWAYS 52(7) Jul '76, 2-13

421 INDIGO
     Pinckney, Elise. "Indigo." AMER DYESTUFF REP
     65(3) Mar '76, 36, 38-39

422 INDUSTRIAL ARTS
     Herschbach, Dennis R. "The origin and mean-
     ing of industrial arts." MAN/SOC/TECHNOL 35
     (5) Feb '76, 134-37 (18 ref.)

423 INDUSTRIAL ARTS--FORECASTS
     Mangano, Ronald M. "Industrial arts, tech-
     nology, and the future." MAN/SOC/TECHNOL
     35(5) Feb '76, 141, 147-49, 158 (4 ref.)

424 INDUSTRIAL ARTS EDUCATION
     McPherson, William H. "Industrial arts edu-
     cation identity crisis." MAN/SOC/TECHNOL
     35(5) Feb '76, 138-40 (14 ref.)

425 INDUSTRIAL ENGINEERING
     Parks, George M., and Roger B. Collins.
     "200 years of industrial engineering." INDUST
     ENG 8(7) Jul '76, 14-25 (28 ref.)

426 INDUSTRIAL MEDICINE
     Felton, Jean S. "200 years of occupational
     medicine in the U.S." J OCCUP MED 18(12) Dec
     '76, 809-17 (77 ref.)

427 INDUSTRIAL NURSING--FORECASTS
     Saller, Dorothy. "Nursing in industry." RN
     39(6) Jun '76, 25, 28, 32

428 INDUSTRY--FORECASTS
     Henry, Richard J. "What is private industry's
     role in America's future?" INDUST WEEK 190(1)
     Jul 5, '76, 67-70

     INFANTILE PARALYSIS. See POLIOMYELITIS--THERA-
     PY 721

429   INFANTS--MORTALITY
        Romanofsky, Peter. "Infant mortality, Dr.
        Henry Dwight Chapin, and the Speedwell So-
        ciety, 1890-1920." MED SOC NJ J 73(1) Jan
        '76, 33-38 (37 ref.)
430   INFLATION (FINANCE)--WATERWORKS
        Abbott, James B. "Inflation must be held in
        check." WATER WASTES ENG 13(7) Jul '76, 68,
        70
431   INFLUENZA
        Falk, Victor S. "The influenza epidemic of
        1918." WIS MED J 75(8) Aug '76, 31-34
432   INFORMATION SCIENCE
        Adkinson, Burton W. "Federal governments sup-
        port of information activities." AMER SOC
        INFOR SCI BULL 2(8) Mar '76, 24-26
433   -----
        Becker, Joseph. "The rich heritage of infor-
        mation science." AMER SOC INFORM SCI BULL
        2(8) Mar '76, 9-13
434   -----
        Burchinal, Lee G. "Bring the American Revolu-
        tion on-line-information science and national
        R and D." AMER SOC INFORM SCI BULL 2(8) Mar
        '76, 27-28
435   -----
        Meadow, Charles T. "1776 and all that." AMER
        SOC INFORM SCI BULL 2(8) Mar '76, 7-8
436   -----
        Mooers, Calvin N. "Technology of information
        handling--a pioneer's view." AMER SOC INFORM
        SCI BULL 2(8) Mar '76, 18-19

437 -----
   Saracevic, Tefko. "Intellectual organization
   of knowledge: the American contribution."
   AMER SOC INFORM SCI BULL 2(8) Mar '76, 16-17
438 -----
   Zurkowski, Paul G. "Perspectives on the information industry." AMER SOC INFORM SCI BULL
   2(8) Mar '76, 28-29
439 INFORMATION SCIENCE--BIBLIOGRAPHY
   Belzer, Jack, and C. Kamila Robertson. "Key
   publications in information science--a selected list." AMER SOC INFORM SCI BULL 2(8)
   Mar '76, 35-38 (144 ref.)
440 INFORMATION SCIENCE--CHRONOLOGY
   Emard, Jean Paul. "An information science
   chronology in perspective." AMER SOC INFORM
   SCI BULL 2(8) Mar '76, 51-56
441 INFORMATION SCIENCE--FORECASTS
   Namus, Burt. "Information Science and the
   future." AMER SOC INFORM SCI BULL 2(8) Mar
   2(8) Mar '76, 57-58
442 -----
   Wooster, Harold A. "The American information
   science society--2076." AMER SOC INFORM SCI
   BULL 2(8) Mar '76, 59-60
443 INFORMATION SCIENCE--PICTORIAL WORKS
   "Faces and facades." AMER SOC INFORM SCI
   BULL 2(8) Mar '76, 42-43
444 INFORMATION SCIENCE EDUCATION
   Taylor, Robert S. "On education." AMER SOC
   INFORM SCI BULL 2(8) Mar '76, 34
445 INSTITUTE OF ELECTRICAL AND ELECTRONICS EN-
   GINEERS

Coggeshall, Ivan S. "Ten vignettes of an
engineering institute." IEEE PROC 64(9) Sep
'76, 1392-99

446 INSTITUTE OF RADIO ENGINEERS
Layton, Edwin T., Jr. "Scientists and engineers: the evolution of the IRE." IEEE PROC
64(9) Sep '76, 1390-92 (14 ref.)

INSTRUMENT MANUFACTURE. See SURVEYING 856

447 INTEGRATED CIRCUITS
"The integrated circuit era (1959-1975) com-
pressing the world of electronics." ELEC-
TRONIC DESIGN 24(4) Feb 16, '76, 118-27

448 -----
Wolff, Michael F. "The genesis of the integrated circuit." IEEE SPECTRUM 13(8) Aug '76,
44-53

449 INTERNAL COMBUSTION ENGINE
Cummins, C. Lyle, Jr. "The first century of
the Otto engine." AUTOMOTIVE ENG 84(7) Jul
'76, 36-45

450 -----
"The Otto-cycle engine: force for change."
AUTOMOTIVE ENG 84(7) Jul '76, 46-51

451 INTERNAL MEDICINE
Petersdorf, R. G. "Internal medicine 1976:
consequences of subspecialization and technology." ANN INTERN MED 84(1) Jan '76, 92-94
(5 ref.)

452 INTERNAL MEDICINE--NEW YORK
Berger, Herbert. "Internal medicine." NY
STATE J MED 76(7) Jul '76, 1184-87

453 INVENTIONS
Danilov, Victor J. "The Bicentennial 100,

America's greatest discoveries, inventions,
and innovations." <u>IND</u> <u>RES</u> 18(12) Nov 15, '76,
21-24

IRON INDUSTRY. See STEEL INDUSTRY 841-43

454 IROQUOIS

"The Iroquois of New York-their past and
present." <u>CONSERVATIONIST</u> 30(4) Jan-Feb '76:
The Hodinonshonni. Lloyd M. Elm, Sr.; Archa-
eology of the Iroquois. Donald L. Tuttle;
All people, all tribes, all nations. Robert
E. Powless; From Lake Champlain to Eagle Bay.
Philip H. Tarbell; Separate yet sharing. Ann
Lewis; The feast. Theodore C. Williams (Hro-
os'ska(t)); A portfolio of Iroquois art and
craft. Intro. by Gerald Pete Jamison; We can
never go back to the woods again. Dale White;
Prehistoric Iroquois medicine. Richard E.
Hosbach and Robert E. Doyle; The Six Nations
and the state. Martin Wasser; Inspired by
Indians. William H. Carr; North American In-
dian traveling college

455 JACKSON, HALL

Estes, J. Worth. "'A disagreeable and danger-
ous employment': medical letters from the
Siege of Boston, 1775." <u>J</u> <u>HIST</u> <u>MED</u> 31(3)
Jul '76, 271-91 (60 ref.)

456 JAMESTOWN, VIRGINIA--COLONIAL PERIOD

"Jamestown, dawn of a new era." <u>NAT</u> <u>PARKS</u> <u>&</u>
<u>CON</u> <u>MAG</u> 50(4) Apr '76, 10-12

457 JAYNE, GERSHON

Barringer, Floyd S. "Pioneer physicians in
Illinois-Gershon Jayne, M.D. (1791-1867)/
Springfield." <u>ILL</u> <u>MED</u> <u>J</u> 150(1) Jul '76,

54-55, 86 (9 ref.)

458 JEFFERSON, THOMAS

Adcock, Louis H. "Thomas Jefferson, scientist." CHEMISTRY 48(8) Sep '75, 14-15

459 -----

Coonen, Lester P., and Charlotte M. Porter.
"Thomas Jefferson and American biology."
BIOSCIENCE 26(12) Dec '76, 745-50 (13 ref.)

460 -----

Ewan, Joseph. "How many botany books did
Thomas Jefferson own?" MO BOT GARD BULL 64
(6) Jun '76, 5-12

461 -----

Gillette, David D. "Thomas Jefferson's pur-
suit of illusory fauna." FRONTIERS 40(3)
Spr '76, 16-21

462 -----

Mitchell, Henry. "Thomas Jefferson, the young
gardener." HORTICULTURE 54(6) Jun '76, 38-51

463 -----

Newbold, Vern. "Thomas Jefferson the sheepman." SHEEP BREED SHEEPMAN 96(5) May '76,
58-60, 62

464 -----

Peterson, Merrill D. "Thomas Jefferson and
the revolution of the mind." ARCH DERM 112
(Spec. Issue) Nov 29, '76, 1637-41

465 -----

Soldati, Gary D. "Jefferson, Clark, and 'the
year of blood'--a search for fossils in the
midst of the Revolution." AMER DENT ASSN J
93(1) Jul '76, 76-77

466   -----

   Spratt, John S. "Thomas Jefferson: the schol-
   ary politician and his influence on medicine." SOUTH MED J 69(3) Mar '76, 360-66
   (29 ref.)

467   -----

   "Thomas Jefferson: the sheepman." NAT WOOL
   GROW 66(4) Apr '76, 10-11, 24-25

468   -----

   Watson, Francis J. B. "America's first uni-
   versal man had a very acute eye." SMITHSON-
   IAN 7(3) Jun '76, 88-90, 92-95

469   -----

   WEAVER, NEAL. "Thomas Jefferson, statesman,
   artist, scientist and one man horticultural
   exchange." GARDEN J 26(5) Oct '76, 147-150
   -----. See also MONTICELLO 607

JEFFERSON, THOMAS--ANECDOTES, FACETIAE, SATIRE,
   ETC. See FRANKLIN, BENJAMIN--ANECDOTES, FACE-
   TIAE, SATIRE, ETC. 344

470   JEWISH CHILDREN--COLONIAL PERIOD

   Bloch, Harry. "Jewish children in Colonial
   times." AMER J DIS CHILD 130(7) Jul '76, 711-
   13 (45 ref.)

471   JEWS--MEDICINE

   Rosner, Fred. "Jewish contributions to medicine in United States, 1776-1976." NY STATE
   J MED 76(8) Aug '76, 1327-32 (23 ref.)

472   JONES, JOHN

   Stark, Richard B. "John Jones, M.D., 1729-
   1791, father of American surgery." NY STATE
   J MED 76(8) Aug '76, 1333-38 (4 ref.)

JONES, JOHN PAUL. See BON HOMME RICHARD 107

473 KHAN, FAZLUR
     Rigoni, Donald L., Jr. "Fazlur Khan: a look
     to the sky." CIVIL ENG-ASCE 46(7) Jul '76, 69
474 KINNEY, J. P.
     Carr, William H. "Three distinguished con-
     servationist." AMER FOR 82(6) Jun '76, 8
KIRKBRIDE PLAN. See PSYCHIATRIC HOSPITALS 738
475 LABOR AND LABORING CLASSES
     Thompson, Donald B. "The working American."
     INDUST WEEK 190(1) Jul 5, '76, 47-66
LAFAYETTE, MARIE JOSEPH, MARQUIS de. See TUCK-
ER, JOSHUA 916
476 LAISSEZ-FAIRE
     McCaull, Julian. "Pursuit of property."
     ENVIRONMENT 18(6) Jul-Aug '76, 17-20, 25-31
     (27 ref.)
477 -----
     Smith, F. Michael, Jr., and H. Ashton Thomas.
     "The birthday of free enterprise: William
     Bradford's manuscript." LOUIS STATE MED SOC
     J 128(11) Nov '76, 297-99
LAND RUSH. See PUBLIC LANDS--OKLAHOMA 752
LAND SPECULATION. See GEOGRAPHY, CULTURAL 359
478 LANGLEY, SAMUEL PIERPONT
     "Samuel Pierpont Langley (1834-1906)."
     ASTRONOMY 4(7) Jul '76, 54
479 LATROBE, BENJAMIN
     Rigoni, Donald L., Jr. "Benjamin Latrobe:
     helping cleanse America." CIVIL ENG-ASCE 46
     (7) Jul '76, 80
LEAD MINES AND MINING. See MINES AND MINERAL
RESOURCES--MASSACHUSETTS 604

480 LEAVITT, HENRIETTA SWAN
     Mitchell, Helen Buss. "Henrietta Swan Leavitt
     and Cepheid variables." PHYS TEACH 14(3) Mar
     '76, 162-67 (10 ref.)
481 LEIDY, JOSEPH
     Meyers, Arlen. "Joseph Leidy, his contribu-
     tions to otology." ARCH OTOLARYNGOL 102(9)
     Sep '76, 576-77
482 LENOX HILL (NEW YORK CITY) HOSPITAL
     Berry, Edgar P. "Lenox Hill Hospital, birth
     and growth." NY STATE J MED 76(4) Apr '76,
     586-91 (6 ref.)
483 LEWIS AND CLARK EXPEDITION
     Chuinard, Eldon G. "Lewis and Clark and the
     Bicentennial." AMER MED ASSN J 236(5) Aug 2,
     '76, 496-98
484 -----
     "Lewis and Clark, a nation looks west." NAT
     PARK & CON MAG 50(9) Sep '76, 11-13
     -----. See also AMERICA--DISCOVERY AND EXPLOR-
     ATION 25
485 LEWIS AND CLARK EXPEDITION--PICTORIAL WORKS
     Satterfield, Archie. "In the footsteps of
     Lewis and Clark." NAT WILDLIFE 14(6) Oct-
     Nov '76, 24-33
486 LIAUTARD, ALEXANDRE
     Crawford, Lester M. "A tribute to Alexandre
     Liautard, the father of the American vet-
     erinary profession." AMER VET MED ASSN J
     169(1) Jul 1, '76, 35-37 (15 ref.)
487 LIBERTY BELL
     "America's Liberty Bell-symbol of freedom."
     FUELOIL OIL HEAT 35(7) Jul '76, 21

488 -----
     Black, Roe C. "How farmers saved the Liberty
     Bell." FARM J 100 (11) Nov '76, G-2
489 -----
     Dieter, Henning B. "Pass and Stow: Liberty
     Bell casters." FOUNDRY MGT & TECH 104(7) Jul
     '76, 36, 38-39
490 -----
     Miller, Eileen. "The Liberty Bell." SCI &
     CHILD 13(4) Jan '76, 31-32
491 LIBERTY BELL--NONDESTRUCTIVE TESTING
     Clarke, E. T. "Radiographing the Liberty
     Bell." FOUNDRY TRADE J 141(3091) Aug 5, '76
     223-24, 226-28
492 -----
     Hanson, Victor F., Janice H. Carlson, Karen
     M. Papouchado, and Norman A. Nielson. "The
     Liberty Bell: composition of the famous
     failure." AMER SCI 64(6) Nov-Dec '76, 615-19
     (10 ref.)
493 -----
     "How they did it--radiographing the Liberty
     Bell." MATERIALS EVALUATION 34(2) Feb '76,
     14A-16A, 18A, 26A
494 -----
     "Proclaim liberty throughout the land . . . "
     ASTM STAND NEWS 4(7) Jul '76, 25-29
495 -----
     "Saving the Liberty Bell through radiography."
     FOUNDRY MGT & TECH 104(7) Jul '76, 26-27, 30,
     32
496 -----
     "X-rays check condition of the Liberty Bell."

IRON AGE 217(7) Feb 16, '76, 33-35

497 LIBRARIES

Shera, Jesse H. "Two centuries of American librarianship." AMER SOC INFORM SCI BULL 2(8) Mar '76, 39-40

498 LICENSURE, MEDICAL

Derbyshire, Robert C. "Medical licensure and professional discipline 1976." ANN INTERN MED 85(3) Sep '76, 384-85 (7 ref.)

499 LICENSURE, MEDICAL--CONNECTICUT--19th CENTURY

"Take this quiz and get a (19th century) license." CONN MED 40(12) Dec '76, 848

500 LICENSURE, MEDICAL--ILLINOIS

Schnepp, Kenneth H. "Medical licensure in Illinois: an historical review." ILL MED J 150(3) Sep '76, 229-34, 248

501 LINEN

Stefanides, E. J. "Making of linen is part of life's fabric." DESIGN N 31(13) Jul 4, '76, 22-23

502 LIVESTOCK

"A history of cattlemen." DROVERS J Aug 9, '76, 21-80 passim

503 LIVESTOCK--FORECASTS

"A long look ahead." DROVERS J Aug 9, '76, 101-76 passim

504 LIVESTOCK--VETERINARY MEDICINE

Kingrey, B. W. "Farm animal practice in the United States." AMER VET MED ASSN J 169(1) Jul 1, '76, 56-60 (15 ref.)

505 LLEWELLYN, DAVID HERBERT

Holley, Howard L. "A century and a half of the history of the life sciences in Alabama."

ALA J MED SCI 13(3) Jul '76, 295

506 LONG, CRAWFORD WILLIAMSON
     "Crawford Williamson Long: his life and his
     discovery." MED ASSN GEORGIA J 65(7) Jul '76,
     280-86

507 LONG ISLAND, BATTLE OF, 1776--MEDICAL ASPECTS
     Shalley, Doris P. "General John Sullivan in
     New Jersey." MED SOC NJ J 73(11) Nov '76,
     983-84 (8 ref.)

508 LONG ISLAND, BATTLE OF, 1776--WEATHER
     Ludlum, David M. "The weather of American
     independence--3: the Battle of Long Island."
     WEATHERWISE 28(3) Jun '75, 118-21, 147

     LONGEVITY. See EMIGRATION AND IMMIGRATION 267

509 LONGEVITY--DECLARATION OF INDEPENDENCE SIGNERS
     "Longevity of signers of the Declaration of
     Independence." METROPOL LIFE STAT BULL 57,
     Apr '76, 2-4

510 LONGEVITY--UNITED STATES--18th CENTURY
     "Longevity in the thirteen original states."
     METROPOL LIFE STAT BULL 57, Feb '76, 2-4

511 LONGEVITY--UNITED STATES--FIRST LADIES
     "Longevity of first ladies of the United
     States." METROPOL LIFE STAT BULL 58, Jan '77,
     2-4

512 LONGEVITY--UNITED STATES--PRESIDENTS
     "Longevity of presidents of the United
     States." METROPOL LIFE STAT BULL 57, Mar '76,
     2-4

513 LOUISIANA PURCHASE
     Fleming, John. "Louisiana Purchase monument."
     AMER CONGR SURV MAP BULL (54) Aug '76, 7-9

LOWE, THADDEUS S.C. See UNITED STATES--CIVIL
WAR 945

514 LOYD, SAM
Spaulding, Raymond E. "Sam Loyd, America's
greatest puzzlist." MATH TEACHER 69(3) Mar
'76, 201-11 (1 ref.)

515 MC CALL'S (PERIODICAL)
100th anniversary issue. MC CALL'S, 103(7)
Apr '76. Contents: The biggest, the best,
the first, and the worst: a capsule history
of the century. Vivian Cadden and Helen
Markel; The pioneer spirit. Lady Bird Johnson; "When I was young, 100 years ago. . .";
The triumph of Marian Anderson. Margaret
Truman; Home sweet home, 1876. Suzanne Hil-
ton; The first woman president. Art Buch-
wald; Marriage and family: from popping the
question to the pill. Margaret Mead; Manners
and mores: the importance of the right fork.
Charlotte Curtis; Politics: what have women
really won?, Frances Fitzgerald; Careers:
when women will be superior to men. Clare
Booth Luce; Heros and villains: who was
famous and why. Jean Stafford; Salvation:
from Billy Sunday to the Beatles. Jessamyn
West

516 MC CARTHY, CLEM
Hagen, Chet. "Matchless Clem McCarthy."
HORSEMEN'S J 27(2) Feb '76, 32-34

517 Mc Clelland, William
"William McClelland, M.D., 1768-1812, first
president of the Medical Society of the State
of New York--1807." NY STATE J MED 76(7)

Jul '76, 1156-59

518 MACHINE TOOLS

"The machine tools that are building America." IRON AGE 218(9) Aug 30, '76, 155-292 passim

519 MACHINERY

Braverman, Harry. "Control of the machine." ENVIRONMENT 18(7) Sep '76, 26-36 (19 ref.)

520 -----

"The men and machines that made America." IRON AGE 217(15) Apr 12, '76, 113-259 passim

521 -----

Rhea, Nolan W. "When American industry came of age." TOOL PROD 42(4) Jul '76, 44-45

522 MAC KAYE, BENTON

Oehser, Paul H. "Three distinguished conservationists." AMER FOR 82(6) Jun '76, 10-11

MAGNETISM

See MAYER, ALFRED MARSHALL 535

523 MALARIA

"The Agues-Pennsylvania." MORBID MORTAL WEEK REP 25(25, pt. 2) Jul 2, '76, 2 (1 ref.)

524 MAN--GENETIC RESEARCH

Pines, Maya. "Genetic profiles will put our health in our own hands." SMITHSONIAN 7(4) Jul '76, 86-91

525 MAN--PREHISTORIC--NEW JERSEY

Aiello, Lucy L. "A 19th century odyssey." MED SOC NJ J 73(1) Jan '76, 23-26 (12 ref.)

526 MANHATTAN EYE, EAR AND THROAT HOSPITAL (NEW YORK CITY)

Rogers, Blair O. "Manhattan Eye, Ear and

Throat Hospital, history and contributions."
NY STATE J MED 76(9) Sep '76, 1555-62 (19
ref.)

527 MARINE ENGINEERING

Hooper, Edwin B. "Over the span of 200
years--technology and the United States
Navy." NAVAL ENG J 88(4) Aug '76, 17-23

528 MARS (PLANET)

Klein, Harold P., Joshua Lederberg, Alexander Rich, Norman H. Horowitz, Vance I. Oyama,
and Gilbert V. Levin. "The Viking Mission
search for life on Mars." NATURE 262(5563)
Jul 1, '76, 24-27 (16 ref.)

529 -----

Zimmerman, Mark, D. "U.S. technology reaches
Mars." MACHINE DESIGN 48(15) Jun 24, '76,
18-22, 24-26

530 MARSHALL, HUMPHRY

Parker, Laura. "Humphry Marshall (1722-1801)
author of the first American book on trees."
MORTON ARBOR Q 12(1) Spr '76, 10-11

531 MARYLAND STATE MEDICAL JOURNAL (PERIODICAL)

Miles, Lester H. "The Med-Chi Journal his-
tory--1839-1976." MD STATE MED J 25(1) Jan
'76, 35-42

MASTS AND RIGGINGS--GREAT BRITAIN--NAVY--
COLONIAL PERIOD. See UNITED STATES--COLONIAL
PERIOD 919

532 MATERIALS--TRANSPORTATION

"Automotive achievement: moving material."
AUTOMOT IND 155(1) Jul 1, '76, 78-79

533 MATHEMATICS

Frisinger, H. Howard. "Mathematics and our

founding fathers." <u>MATH</u> <u>TEACHER</u> 69(4) Apr '76
301-17 (18 ref.)

534 -----

Jones, Phillip S. "1876±100." <u>MATH</u> <u>TEACHER</u>
69(7) Nov '76, 586-93 (5 ref.)

535 MAYER, ALFRED MARSHALL

Snelders, H. A. M. "A. M. Mayer's experiments
with floating magnets and their use in the
atomic theories of matter." <u>ANN</u> <u>SCI</u> 33(1)
Jan '76, 67-80 (64 ref.)

536 MEASLES

"Measles--Small Pokkes on shipboard." <u>MORBID</u>
<u>MORTAL</u> <u>WEEK</u> <u>REP</u> 25(25 pt. 2) Jul 2, '76, 3

537 MECHANICAL ENGINEERING

"200 years of mechanical engineering." <u>MECH</u>
<u>ENG</u> 98(1) Jan '76, 29-33

538 MEDICAL CARE

Dock, William. "Medical care, two centuries
of revolutionary change." <u>NY</u> <u>STATE</u> <u>J</u> <u>MED</u>
76(7) Jul '76, 1121-22

539 -----

Edwards, Ralph. "Health and medical care at
the time of the American Revolution." <u>J</u> <u>SCH</u>
<u>HEALTH</u> 46(1) Jan '76, 19-23 (8 ref.); (2)
Feb '76, 77-80 (8 ref.)

-----. See also GROUP MEDICAL PRACTICE 372;
HESSIAN MERCENARIES--REVOLUTIONARY WAR 388
MEDICAL CARE--18th CENTURY. See U.S.S. CON-
STITUTION--MEDICAL ASPECTS 991
MEDICAL CARE--DELAWARE--19th CENTURY. See HOMO-
PATHY 390

540 MEDICAL CARE--FORECASTS

Cathcart, H. Robert. "What is ahead for

healthcare?" <u>MOD</u> <u>HEALTHCARE</u> 6(1) Jul '76,
20-22

541 MEDICAL CARE--ILLINOIS--18th CENTURY
Pearson, Emmet F. "Health care in Illinois
circa 1776." <u>ILL</u> <u>MED</u> <u>J</u> 150(6) Dec '76, 608-
11

542 MEDICAL CARE--NEW JERSEY--19th CENTURY
Henderson, Alfred R. "Prominent medicine
convenes at Lumberton, 1887." <u>MED</u> <u>SOC</u> <u>NJ</u> <u>J</u>
73(1) Jan '76, 18-22 (25 ref.)
MEDICAL CARE--NEW YORK. See GENERAL PRACTICE--
NEW YORK 357

543 MEDICAL CARE--OHIO--19th CENTURY
Sidall, A. Claire. "Consumer rebellion
against orthodox medicine at Oberlin--1833."
<u>OHIO</u> <u>STATE</u> <u>MED</u> <u>J</u> 72(6) Jun '76, 330-34 (41
ref.)

544 MEDICAL CARE--RHODE ISLAND--18th CENTURY
Gilman, John F. W. "Aesculapius comes to the
Colonies." <u>RHODE</u> <u>ISLAND</u> <u>MED</u> <u>J</u> 59(7) Jul '76,
328-32, 342-44 (17 ref.)

545 MEDICAL CARE--VERDE VALLEY, ARIZONA
Leyel, Martha, and M. E. Linkert. "A Bicen-
tennial salute to health services in the
Verde Valley of Arizona." <u>ARIZONA</u> <u>MED</u> 33(6)
Jun '76, 500-509

546 MEDICAL CENTERS--NEW YORK
"Medical centers." <u>NY</u> <u>STATE</u> <u>J</u> <u>MED</u> 76(7) Jul
'76, 1219-35, 1238-39

547 MEDICAL EDUCATION
Jarcho, Saul. "The fate of British traditions
in the United States as shown in medical education and in the care of the mentally ill."

NY ACAD MED BULL 52(3) Mar-Apr '76, 419-41
(45 ref.)

548 -----
Lambert, Samuel W. "Changes in medical teaching over the past century." NY ACAD MED BULL
52(3) Mar-Apr '76, 270-77 (3 ref.)

549 -----
Lee, Philip R. "Graduate medical education
1976? Will internal medicine meet the challenge?" ANN INTERN MED 85(2) Aug '76, 250-53
(9 ref.)

MEDICAL EDUCATION, CONTINUING. See SOCIETIES,
MEDICAL--19th CENTURY 831

MEDICAL HISTORY--18 CENTURY. See MIDDLETON,
PETER 596

550 MEDICAL LITERATURE--18th CENTURY
Beatty, William K. and Virginia L. Beatty.
"Sources of medical information." AMER MED
ASSN J 236(1) Jul 5, '76, 78-82 (9 ref.)

551 -----
Blake, John B. "Early American medical lit-
erature." AMER MED ASSN J 236(1) Jul 5, '76,
41-46 (4 ref.)

552 MEDICAL SERVICE, COST OF
Somers, H. M. "Health care 1976: cost and
consequences." ANN INTERN MED 84(2) Feb '76,
211-12 (9 ref.)

553 MEDICAL SOCIETY OF THE STATE OF NEW YORK--
WOMAN'S AUXILIARY
Ulrich, Helen. "Woman's auxiliary, true ally
of The Medical Society." NY STATE J MED 76
(7) Jul '76, 1254-55

MEDICAL STATISTICS. See PSYCHIATRY 740

554   MEDICAL STUDENTS--COLONIAL PERIOD

Goodrich, James T. "The Colonial American medical student: 1750-1776." CONN MED 40(12) Dec '76, 829-44 (139 ref.)

555   MEDICINE

Bean, William B. "Medical practice: past, present, and future." J MED EDUC 51(12) Dec '76, 979-85

556   -----

Blaylock, Russell L. "American Medicine: Bicentennial celebration." SO CAROLINA MED ASSN J 72(1) Jan '76, 18-22 (25 ref.)

557   -----

Greifinger, Robert B., and Victor W. Sidel. "American medicine." ENVIRONMENT 18(4) May '76, 6-18

558   -----

Huth, Edward J. "1976: an agenda for American medicine." ANN INTERN MED 85(6) Dec '76, 818-19

559   -----

Ober, William B., and John H. Edgcomb. "Two centuries of medical progress." NY STATE J MED 76(7) Jul '76, 1106-18

-----. See also BRADLEY, SAMUEL BEACH 115; JEFFERSON, THOMAS 466; JEWS--MEDICINE 471

560   MEDICINE--COLONIAL PERIOD

Caldwell, Hayes W. "Colonial medicine and Benjamin Franklin." ARIZONA MED 33(1) Jan '76, 55-59; (2) Feb '76, 129-33

561   -----

Graney, Charles M. "Colonial medicine." NY

STATE J MED 76(7) Jul '76, 1123-25

MEDICINE--15th CENTURY. See AMERICA--DISCOVERY
AND EXPLORATION 26

562 MEDICINE--18th CENTURY

"American medicine in 1776." KAN MED SOC J
77(7) Jul '76, 347

563 -----

Bloom, Mark. "Battlefield medicine in the
American Revolution." MED WORLD NEWS 17(14)
Jun 28, '76, 43-46, 50, 53-56

564 -----

Bloom, Mark. "Medicine on the eve of the
Revolution." MED WORLD NEWS 16(27) Dec 15,
'75, 45-48, 51, 58-59

565 -----

Davis, Audrey B. "Medicine in the Revolutionary period." SCI & CHILD 13(4) Jan '76,16-18

566 -----

Der Marderosian, Ara and Mukund S. Yelvigi.
"Medicine and drugs in Colonial America."
AMER J PHARM 148(4) Jul-Aug '76, 113-17 (10
ref.)

567 -----

Graney, Charles M. "Medicine in the Revolution." NY STATE J MED 76(7) Jul '76, 1125-30

568 MEDICINE--19th CENTURY

Howard, John W. "Nineteenth century medicine." DEL MED J 48(10) Oct '76, 581-82

569 -----

Kirkwood, Tom. "Medicine a century ago."
ILL MED J 149(4) Apr '76, 370-75 (6 ref.)

-----. See also OSLER, WILLIAM 677

570 MEDICINE--CALGARY, CANADA
     Troy, M. Tuszewski. "Early medicine and sur-
     gery in Calgary." CANAD J SURG 19(5) Sep '76,
     449-53 (2 ref.)

571 MEDICINE--CHARLESTON, SOUTH CAROLINA--18th
     CENTURY
     Waring, Joseph I. "Medicine in Charleston at
     the time of the Revolution." AMER MED ASSN J
     236 (1) Jul 5, '76, 31-34 (11 ref.)

572 MEDICINE--CHARLESTON, SOUTH CAROLINA--19th
     CENTURY
     Waring, Joseph I. "Charleston medicine 1800-
     1860." J HIST MED 31(3) Jul '76, 320-42 (72
     ref.)

573 MEDICINE--DELAWARE, LEWES--19th CENTURY
     Howard, John W. "Nineteenth century medicine
     in Lewes." DEL MED J 48(7) Jul '76, 411-15

574 MEDICINE--ILLINOIS--18th CENTURY
     Malkovich, Daniel. "Some notes on early medi-
     cal practice in Illinois, before 1800."
     ILL MED J 149(1) Jan '76, 43-48
     MEDICINE--ILLINOIS, LAWRENCE CO.--19th CENTURY
     See CATTERTON, DYLER 137

575 MEDICINE--ILLINOIS, LOGAN CO.
     Barringer, Floyd S. "A history of the prac-
     tice of medicine in Logan County, Illinois."
     ILL MED J 149(3) Mar '76, 295-97 (10 ref.)

576 MEDICINE--NEW JERSEY
     Saffron, Morris H. "Remarks on New Jersey's
     contribution to medicine." MED SOC NJ J 73(1)
     Jan '76, 14-15

577 MEDICINE--NEW YORK CITY--18th CENTURY
     Cushman, Paul, Jr. "Revolutionary War; im-

pact on medicine in New York City." NY STATE
J MED 76(9) Sep '76, 1543-51 (32 ref.)

578 MEDICINE--OHIO--19th CENTURY
Hartung, Walter H., and Max T. Schnitker.
"The swamp physician: medical practice in
northwestern Ohio around 1850." OHIO STATE
MED J 72(5) May '76, 317-20 (6 ref.)

579 MEDICINE--PICTORIAL WORKS
"200 years of American medicine." MED TIMES
104(7) Jul '76, 49-56

580 MEDICINE--RHODE ISLAND--18th CENTURY
Millard, Charles E. "Rhode Island medicine
in the Revolution." RHODE ISLAND MED J 59
(7) Jul '76, 299-316, 334, 338-39 (18 ref.)

581 MEDICINE--VIRGINIA--18th CENTURY
Jones, Gordon W. "Medicine in Virginia in
Revolutionary times." J HIST MED 31(3) Jul
'76, 250-70 (45 ref.)

582 MEDICINE--WISCONSIN, DANE CO.--19th CENTURY
Herriott, Marianne. "Placards and pretenders:
the Dane County Medical Society in the 'good
old days'." WIS MED J 75(10) Oct '76, 14-16
(3 ref.)

583 MEDICINE, COMPARATIVE
Bustad, Leo K., John R. Gorham, Gerald A.
Hegreberg, and George A. Padgett. "Compara-
tive medicine: progress and prospects." AMER
VET MED ASSN J 169(1) Jul 1, '76, 90-105
(197 ref.)
-----. See also UNITED STATES--AIR FORCE--VET-
ERINARY MEDICINE 925

584 MEDICINE IN ART
Nathan, Helmuth M. "Art in medicine." NY

STATE J MED 76(7) Jul '76, 1135-40
585   -----
Ravin, James G. "American art and medicine."
OHIO STATE MED J 72(2) Feb. '76, 74-76 (5 ref.)
586   MEDICINES, PATENT, PROPRIETARY, ETC.
Der Marderosian, Ara. "Good for what ails
you!" AMER J PHARM 148(4) Jul-Aug '76, 121-
24 (4 ref.)
587   MENTAL HEALTH--OCCUPATIONS
"The mental health disciplines." HOSP COM-
MUNITY PSYCHIATRY 27(7) Jul '76, 495-504
588   MENTAL HEALTH SERVICES
Brown, Bertram S. "The federal mental health
program: past, present, and future." HOSP
COMMUNITY PSYCHIATRY 27(7) Jul '76, 512-514
589   -----
Dain, Norman. "From Colonial America to Bicentennial America: two centuries of vicis-
situdes in the institutional care of mental
patients." NY ACAD MED BULL 52(10)Dec '76,
1179-96 (21 ref.)
590   MENTAL HEALTH SERVICES--FORECASTS
Resnick, Eugene V. "Mental health care in
America: 2076." HOSP COMMUNITY PSYCHIATRY
27(7) Jul '76, 519-21
MENTAL ILLNESS. See MEDICAL EDUCATION 547
MENTALLY HANDICAPPED. See DIX, DOROTHEA LYNDE
236: HOWE, SAMUEL GRIDLEY 406
591   MERCHANT SHIPS
"18th century tobacco merchant ship recon-
structed." TOBACCO REP 103(7) Jul '76, 26
592   MERINO SHEEP
"The golden fleece tarnished for early sheep

enthusiasts," <u>NAT</u> <u>WOOL</u> <u>GROW</u> 66(7) Jul '76, 13

593 METALS

"How energy planning is changing metal work-
ing." <u>IRON</u> <u>AGE</u> 217(1) Jan 5, '76, 79-86, 91-
98

594 METEORITES

Lange, Erwin F. "The founders of American
meteoritics." <u>METEORITICS</u> 10(3) Sep '75, 241-
52 (19 ref.)

METROPOLITAN HOSPITAL CENTER (NEW YORK CITY).
See NEW YORK MEDICAL COLLEGE 646

595 MEYER, ADOLF

"Adolph Meyer: studying the whole man." <u>HOSP</u>
<u>COMMUNITY</u> <u>PSYCHIATRY</u> 27(7) Jul '76, 492-93

MICROWAVE OVENS. See STOVES 846

596 MIDDLETON, PETER

Bloch, Henry. "Dr. Peter Middleton (?-1781)."
<u>MED</u> <u>SOC</u> <u>NJ</u> <u>J</u> 73(11) Nov '76, 979-81 (16 ref.)

597 MILITARY ART AND SCIENCE

Clark, Paul W. "Early impacts of communications on military doctrine." <u>IEEE</u> <u>PROC</u> 64(9)
Sep '76, 1407-13 (51 ref.)

598 MILITARY ENGINEERING

Johnson, Leland R. "Sword, shovel, and com-
pass." <u>MIL</u> <u>ENG</u> 68(443) May-Jun '76, 159-65

599 -----

Jones, Matthew J. "200 years of military engineering." <u>ENGINEER</u> (<u>FT.</u> <u>BELVOIR</u>) 6(3) Jul-
Sep '76, 36-40; (4) Oct-Dec '76, 22-24

600 MILITARY MEDICINE--RECRUITMENT

Berry, Frank B. "The story of 'The Berry
plan'." <u>NY</u> <u>ACAD</u> <u>MED</u> <u>BULL</u> 52(3) Mar-Apr '76,
278-82

MILWAUKEE. See SMALLPOX--MILWAUKEE--19th CEN-
TURY 829

601 MINERALS

"The importance of minerals in the U.S. econ-
omy--1776-1976." U.S. BUR MINES BULL (667) '75,
MINERAL FACTS AND PROBLEMS 1-34 (21 ref.)

602 MINES AND MINERAL RESOURCES

Hoffman, John. "Mining frontiers--a Bicentennial review, 1776 to 1976." MIN CONG J
62(2) Feb '76, 63-67

603 MINES AND MINERAL RESOURCES--BIOGRAPHY

Guccione, Eugene. "Great men in American
mining." MIN ENG 28(7) Jul '76, 65-72

604 MINES AND MINERAL RESOURCES--MASSACHUSETTS

Marshall, John H. Jr., and Pete J. Dunn.
"The lead mines of Loudville." ROCK MIN 51
(5) Jun '76, 250-55 (7 ref.)

605 MINES AND MINERAL RESOURCES--NEVADA

Batory, Dana M. "100 years of history:
Nevada's Comstock Lode." ROCK MIN 51(7) Sep
'76, 345-49 (6 ref.)

606 MITCHELL, MARIA

"Maria Mitchell (1818-1884)." ASTRONOMY
4(7) Jul '76, 54

MITCHELL, SILAS WEIR. See NEUROLOGY 642

MITCHILL, SAMUEL L. See BLACK, JOSEPH 104

607 MONTICELLO--PICTORIAL WORKS

"Thomas Jefferson's Monticello--America's
first modern house." HOUSE & GARD 148(1)
Jan '76, 48-53

MORGAN, JOHN. See MEDICINE--18th CENTURY 567

MOTHERS--MORTALITY. See STUDY COMMITTEE OF THE
WISCONSIN MATERNAL MORTALITY SURVEY 849

608 MOTION PICTURE CAMERAS
        Di Giulio, Edmund M. "Developments in motion-
        picture camera design and technology-a ten
        year update." SMPTE J 85(7) Jul '76, 481-87
609 -----
        Di Giulio, Edmund M., E. C. Manderfeld, and
        George A. Mitchell. "A historical survey of
        the professional motion-picture camera."
        SMPTE J 85(7) Jul '76, 487-92 (7 ref.)
610 MOTION PICTURE FILM
        Fordyce, Charles R. "Motion-picture film sup-
        port: 1889-1976, a historical review." SMPTE
        J 85(7) Jul '76, 493-95 (11 ref.)
611 MOTION PICTURE INDUSTRY
        Richardson, F. H. "What happened in the be-
        ginning." SMPTE J 85(7) Jul '76, 572, 574,
        576, 578, 580, 582, 584, 586, 590
612 MOTION PICTURE INDUSTRY--CANADA
        Graham, Gerald G. "The early years of the
        Canadian film industry." SMPTE J 85(7) Jul
        '76, 546-47
613 MOTION PICTURE MACHINES
        Kloepfel, Don V. "Developments in the design
        of projection equipment." SMPTE J 85(7) Jul
        '76, 538-45
614 MOTION PICTURE PHOTOGRAPHY--PRINTING PROCESSES
        Solow, Sidney P. "Milestones in the history
        of the motion-picture film laboratory."
        SMPTE J 85(7) Jul '76, 505-14 (57 ref.)
615 MOTION PICTURE PHOTOGRAPHY--STANDARDS
        Chambers, Gordon A. "A short history of
        standardization in the SMPTE." SMPTE J 85(7)
        Jul '76, 451

616 MOTION PICTURE SOUND RECORDING
     Frayne, John G., Arthur C. Blaney, George R.
     Groves, and Harry F. Olson." A short history
     of motion-picture sound recording in the
     United States." SMPTE J 85(7) Jul '76, 515-
     28 (55 ref.)

617 MOTION PICTURE SOUND RECORDING--EUROPE
     Wohlrab, Hans Chr. "Highlights of the his-
     tory of sound recording on film in Europe."
     SMPTE J 85(7) Jul '76, 531-33 (4 ref.)

618 MOTION PICTURE SOUND RECORDING--MAGNETIC RE-
     CORDING
     Ryder, Loren L. "Magnetic sound recording in
     the motion-picture and television indus-
     tries." SMPTE J 85(7) Jul '76, 528-30

619 MOTION PICTURE STUDIOS--LIGHTING
     Aron, Daniel L. "Advancements in motion-pic-
     ture and television set lighting equipment."
     SMPTE J 85(7) Jul '76, 534-37 (29 ref.)

620 MOTION PICTURES, COLORED
     Ryan, Roderick T. "Color in the motion-pic-
     ture industry." SMPTE J 85(7) Jul '76, 496-
     504 (93 ref.)

621 MOTOR TRUCK INDUSTRY
     Lyndall, Jack. "Heritage '76: from humble
     beginnings." FLEET OWNER 71(1) Jan '76, 50-
     54

622 MOTOR TRUCK INDUSTRY--EQUIPMENT
     Lyndall, Jack. "Equipment: made to do the
     job." FLEET OWNER 71(1) Jan '76, 55-60

623 MOTOR TRUCK INDUSTRY--FORECASTS
     Boyce, Carroll W. "What may 'trucking' be
     like 25 years down the pike." FLEET OWNER

71(1) Jan '76, 75-77

624 MOTOR TRUCK INDUSTRY--LAWS AND REGULATIONS
Byczynski, Stu. "The regulations horizon."
FLEET OWNER 71(1) Jan '76, 70-74

625 MOTOR TRUCK INDUSTRY--PRODUCTIVITY
Swart, Bernie. "More industry productivity:
from management, from materials." FLEET
OWNER 71(1) Jan '76, 61-69

626 MOTOR VEHICLES, MILITARY
"Automotive achievement: military support."
AUTOMOT IND 155(1) Jul 1, '76, 82-83

627 MOTOR VEHICLES, OFF-HIGHWAY
"Automotive achievement: off-highway." AUTO-
MOT IND 155(1) Jul 1, '76, 72-73

628 MOUNT VERNON--PICTORIAL WORKS
"The heritage: Mount Vernon--a private view."
HOUSE & GARD 148(7) Jul '76, 40-47

629 MULES
Ross, Irwin. "Toast to a vanishing American:
the mule." FARM J 100(5) Apr '76, B-6

630 MUNSON, ENEAS
Quen, Jacques M. "Dr. Eneas Munson (1734-
1826)." J HIST MED 31(3) Jul '76, 307-19
(54 ref.)

631 MUSEUM OF SCIENCE AND INDUSTRY (CHICAGO)
"America's inventive genius, Museum of Sci-
ence and Industry, Chicago, 1975-77." APP
OPTICS 15(7) Jul '76, 1706

632 NATIONAL FOUNDATION FOR INFANTILE PARALYSIS
"The National Foundation for Infantile Paralysis." PHYS THER 56(1) Jan '76, 44-46 (5
ref.)

633 NATIONAL LIBRARIES
     Chartrand, Robert L., "Three national li-
     braries." AMER SOC INFORM SCI BULL 2(8) Mar
     '76, 41

634 NATIONAL PARKS
     Sax, Joseph L. "American's national parks,
     their principles, purposes, and prospects."
     NAT HIST 85(8) Oct '76, 57-88

     NATURAL RESOURCES. See HUMAN ECOLOGY 410

635 NAVAL HISTORY--CIVIL WAR
     Ashkenazy, Irvin. "War at sea, 1862." OCEANS
     9(1) Jan-Feb '76, 24-31

636 NAVAL HISTORY--REVOLUTIONARY WAR
     Chapman, R.L. "1775 was a long time ago. In
     200 years . . . we've come a long way!"
     FATHOM 7(2) Fal '75, 1-7

637 -----
     Schuessler, Raymond. "War at Sea, 1776."
     OCEANS 9(1) Jan-Feb '76, 10-15

638 NAVAL HISTORY--WORLD WAR, 1939-45
     Hunt, Jim. "War at sea, 1941." OCEANS 9(1)
     Jan-Feb '76, 38-47

639 NAVIGATION
     Dewey, William T. "Air navigation-the early
     years." NAVIGATOR 23(1) Spr 76, 5-9

640 -----
     Haney, David J. "Navigation north of seventy."
     NAVIGATOR 23(2) Sum '76, 5-10

641 NEEDLEWORK--PICTORIAL WORKS
     "The American story." MOD MATURITY 19(3)
     Jun-Jul '76, 35-38

642 NEUROLOGY
     Spillane, John D. "Hughlings Jackson's

American contemporaries: the birth of American neurology." ROY SOC MED PROC 69(6) Jun '76, 393-408 (84 ref.)

NEUROSURGERY. See CUSHING, HARVEY 215

643 NEUROSURGERY--NEW YORK

McDonald, Joseph V. "History of neurologic surgery in New York State." NY STATE J MED 76(8) Aug '76, 1342-44 (36 ref.)

644 NEW FRANCE

"New France in the new world." NAT PARKS & CON MAG 50(3) Mar '76, 15-17

645 NEW YORK CITY--BATTLE OF, 1776--WEATHER

Ludlum, David M. "The weather of American independence--4: the loss of New York City and New Jersey." WEATHERWISE 28(4) Aug '75, 172-76

NEW YORK CITY-MEDICAL ASPECTS. See CORONERS AND MEDICAL EXAMINERS--NEW YORK CITY 209

646 NEW YORK MEDICAL COLLEGE

Rhoads, Harmon T., and Joseph B. Cleary. "New York Medical College and Metropolitan Hospital Center, one hundred-year affiliation in medical care." NY STATE J MED 76(4) Apr '76, 594-96 (3 ref.)

647 NEW YORK STATE JOURNAL OF MEDICINE (PERIODICAL)

Smith, Elizabeth C. "Our diamond jubilee." NY STATE J MED 76(7) Jul '76, 1147-55

648 NEWELL, WILLIAM AUGUSTUS

Rogers, Fred B. "William Augustus Newell: physician, governor, congressman." MED SOC NJ J 73(12) Dec '76, 1109

NIAGARA FALLS. See ELECTRIC POWER PLANTS-- NIAGARA FALLS 251

649 NIAGARA FALLS--PHOTOGRAPHY
     Jones, Emery. "100 years of power, an his-
     toric look through the Niagara Mohawk files."
     INDUST PHOT 25(7) Jul '76, 35-37
650 NICKERSON, L. H. A.
     "Pioneer physicians in Illinois--L. H. A.
     Nickerson, M.D. (1851-1939)." ILL MED J 150
     (4) Oct '76, 448
651 NONDESTRUCTIVE TESTING--FORECASTS
     Posakony, Gerald J. "A look to the past--a
     look to the future." MATERIALS EVALUATION
     34(12) Dec '76, 10A, 12A, 14A, 16A, 18A-21A
     (18 ref.)
652 NURSES--CIVIL WAR
     Kalisch, Philip A., and Beatrice J. Kalisch.
     "Untrained but undaunted: the women nurses
     of the blue and the gray." NURS FORUM 15(1)
     '76, 4-33 (46 ref.)
653 NURSING
     Ingles, Thelma. "The physicians' view of the
     evolving nursing profession, 1873-1913."
     NURS FORUM 15(2) '76, 123-64 (30 ref.)
654 NURSING--FORECASTS
     Isler, Charlotte. "The promises in your fu-
     ture." RN 39(12) Dec '76, 15-16, 20, 25, 28
655 NURSING--LAWS AND REGULATIONS--FORECASTS
     Regan, William A. "Nursing and the law."
     RN 39(1) Jan '76, 21, 23, 26, 30
656 NURSING--POLITICS AND GOVERNMENT--FORECASTS
     Mullane, Mary K. "Politics begins at work."
     RN 39(7) Jul '76, 45-51
657 NURSING EDUCATION--FORECASTS
     Lenburg, Carrie B. "Nursing and education."

$\underline{RN}$ 39 (3) Mar '76, 21-22, 26, 28, 30

658   NURSING RESEARCH--FORECASTS

Notter, Lucille E. "Nursing and research."
$\underline{RN}$ 39(5) May '76, 19, 22, 24-26

659   NUTRITION

Darby, William J. "Nutrition science: an
overview of American genius." $\underline{NUT}$ $\underline{R}$ 34(1)
Jan '76, 1-14

660   -----

Darby, William J. "The science of nutrition
1776-1976." $\underline{UTAH}$ $\underline{SCI}$ 37(3) Sep '76, 70-74

661   -----

Dwyer, J. T., and M. E. Molitch. "Clinical
nutrition 1976: where are we going?" $\underline{ANN}$
$\underline{INTERN}$ $\underline{MED}$ 84(3) Mar '76, 329-331 (18 ref.)
-----. See also FOOD SUPPLY--18th CENTURY 321

662   NUTRITION--CHRONOLOGY

Todhunter, E. Neige. "Chronology of some
events in the development and application of
the science of nutrition." $\underline{NUT}$ $\underline{R}$ 34(12) Dec
'76, 353-65

663   NUTRITION RESEARCH--18th CENTURY

Bing, Franklin C. "Nutrition research and
education in the age of Franklin." $\underline{AMER}$
$\underline{DIETET}$ $\underline{ASSN}$ $\underline{J}$ 68(1) Jan '76, 14-21 (24 ref.)
OBERLIN COLLEGE--MEDICAL CARE. See MEDICAL
CARE--OHIO--19th CENTURY 543

664   OBSTACLES (MILITARY SCIENCE)

Diehl, William J., Jr. "The great chain,
link to the past." $\underline{SOLDIERS}$ 31(7) Jul '76, 12

665   -----

"The great chain across the Hudson." $\underline{MIL}$ $\underline{ENG}$
68(446) Nov-Dec '76, 441

666 OBSTETRICAL NURSING--FORECASTS
     Bland, Barbara. "OBG nursing." RN 39(4) Apr
     '76, 21, 24-25
     OBSTETRICS--COMPLICATIONS. See PREGNANCY COM-
     PLICATIONS 731
667 OBSTETRICS--NEW YORK
     Hughes, Edward C. "Bicentennial history of
     obstetrics and gynecology in New York State."
     NY STATE J MED 76(10) Oct '76, 1890-92 (3
     ref.)
     OCEANOGRAPHIC RESEARCH--18th CENTURY. See
     DE BRAHM, WILLIAM GERARD 220
668 OCEANOGRAPHIC RESEARCH SHIPS
     Nelson, Stewart B. "A Bicentennial retro-
     spective: oceanographic ships in the United
     States." MARINE TECH SOC J 10(6) Jul-Aug '76,
     20-28
     OLD IRONSIDES. See U.S.S. CONSTITUTION--MEDI-
     CAL ASPECTS 991
669 OPHTHALMOLOGY--NEW YORK
     Kara, Gerald B. "Two hundred years of ophth-
     almology in New York State." NY STATE J MED
     76(7) Jul '76, 1187-92 (11 ref.)
670 OPTOMETRY--ANECDOTES, FACETIAE, SATIRE
     "Optical weekly--July 4, 1776." OPTOMETRIC
     WEEKLY 67(31) July 29, '76. Contents:
     Rocky voyage offers little in the way of
     haute cuisine; What price, battle?; Frames,
     lenses et al; Philadelphia news opinion;
     News briefs; Letters to the editor; The ob-
     servations of Dr. William Porterfield.
     Joseph Priestley; Important readings; A new
     method of curing cataract by extraction of

the lens. Jacques Daviel; In praise of optics. William Emerson; Of the gutta serena.
Richard Mead; Commentaries on the history
and cure of visual diseases

671 OPTOMETRY--MARYLAND
Dvorine, Israel. "The early history of op-
tometry in Maryland." AMER OPTOMET ASSN J 47
(12) Dec '76, 1558-67

672 ORPHANS AND ORPHAN ASYLUMS--18th CENTURY
Radbill, Samuel X. "Reared in adversity: institutional care of children in the 18th century." AMER J DIS CHILD 130(7) Jul '76,
751-61 (29 ref.)

673 ORTHOPEDIC NURSING--FORECASTS
Donohoo, Claro, "Orthopedic nursing." RN 39
(10)Oct '76, 19-20, 24, 26-27

674 ORTHOPEDIC SURGERY
Bick, Edgar M. "American orthopedic surgery:
the first 200 years." NY ACAD MED BULL 52(3)
Mar-Apr '76, 293-325

675 -----
Bick, Edgar M. "American orthopedic surgery,
the first 200 years." NY STATE J MED 76(7)
Jul '76, 1192-97; (8) Aug '76, 1346-54 (2
ref.)

ORTHOPEDICS--SAN FRANCISCO. See ORTHOPEDICS--
WESTERN UNITED STATES 676

676 ORTHOPEDICS--WESTERN UNITED STATES
Mandell, Peter, Thomas Raih, and Lloyd W.
Taylor. "A history of orthopedics in San
Francisco and the West." WEST J MED 125(6)
Dec '76, 502-8 (16 ref.)

677 OSLER, WILLIAM
Harvey, A. McGehee. "William Osler and medicine in America: with special reference to the Baltimore period." MD STATE MED J 25(10) Oct '76, 35-42 (8 ref.)
OTOLOGY. See LEIDY, JOSEPH 481
OTTO ENGINE. See INTERNAL COMBUSTION ENGINE 449-50

678 PACK, ARTHUR NEWTON
Carr, William H. "Three distinguished conservationists." AMER FOR 82(6) Jun '76, 8-9

679 PAGE, CHARLES G.
Post, Robert C. "Stray sparks from the induction coil: the Volta prize and the Page patent." IEEE PROC 64(9) Sep '76, 1279-86 (41 ref.)

680 PAINT INDUSTRY AND TRADE--18th CENTURY
Lipper, Cathy. "Colonial paintmakers put down early roots." MOD PAINT COAT 66(7) Jul '76, 35-38
PALEONTOLOGY. See JEFFERSON, THOMAS 461, 465
PANAMA CANAL. See GORGAS, WILLIAM CRAWFORD 368

681 PANAMA RAILROAD
Ebel, Wil. "The Panama railroad, predecessor to the canal." TRANSLOG 7(10) Nov '76, 14-15, 18, 20

682 PAPER INDUSTRY--18th CENTURY
Smith, David C. "The state of the paper industry in 1776." TAPPI 59(7) Jul '76, 56-59 (8 ref.) and correction (12) Dec '76, 131

683 PAPER MAKING
"Paper, a witness to independence." PAP TRADE J 160(13) Jul 1, '76, 37-39

PASS AND STOW (COMPANY). See LIBERTY BELL 489

684 PATHOLOGY--19th CENTURY

Ober, William E. "American pathology in the 19th century: notes for the definition of a specialty." NY ACAD MED BULL 52(3) Mar-Apr '76, 326-47 (26 ref.)

685 PEDIATRIC NURSING--FORECASTS

Barnsteiner, James H. "Pediatric nursing." RN 39(11) Nov '76, 21, 24, 28-29, 31, 32

686 PEDIATRICS

Cone, Thomas E. "Highlights of two centuries of American pediatrics, 1776-1976." J DIS CHILD 130(7) Jul '76, 762-75 (110 ref.)

687 PEDIATRICS--18th CENTURY

Radbill, Samuel X. "Colonial pediatrics." J PEDIATRICS 89(1) Jul '76, 3-7 (19 ref.)

688 -----

Waring, Joseph I. "American pediatric writings of the 18th century." J DIS CHILD 130(7) Jul '76, 741-46 (104 ref.)

689 PEDIATRICS--NEW YORK

Wheatley, George M. "Brief history of pediatrics in New York." NY STATE J MED 76(7) Jul '76, 1197-1201

PELL, THOMAS. See SURGEONS--CONNECTICUT 852

PENNSYLVANIA HOSPITAL. See PHARMACISTS 691

690 PERIODICALS--PSYCHIATRIC

"Asylum periodicals: something new under the sun." HOSP COMMUNITY PSYCHIATRY 27(7) Jul '76, 482-83

691 PHARMACISTS

Williams, William H. "Pharmacists at America's first hospital, 1752-1841." AMER J HOSP

PHARM 33(8) Aug '76, 804-7 (24 ref.)

692 PHARMACOLOGY

Dragstedt, Carl A. "Sidelights of American pharmacology." ILL MED J 150(2) Aug '76, 143-47

693 PHARMACY--18th CENTURY

Cowen, David L. "The foundations of pharmacy in the United States." AMER MED ASSN J 236 (1) Jul 5, '76, 83-87

694 PHARMACY--EDUCATION

Mrtek, Robert G., "Pharmaceutical education in these United States--an interpretive historical essay of the twentieth century." AMER J PHARMACEUT ED 40(4) Nov '76, 339-65 (190 ref.)

695 PHASE RULE AND EQUILIBRIUM

Daub, Edward E. "Gibbs phase rule: a Centenary retrospect." J CHEM ED 53(12) Dec '76, 747-51 (42 ref.)

-----. See also GIBBS, JOSIAH WILLARD 365

696 PHILADELPHIA COLLEGE OF PHARMACY AND SCIENCE

Kramer, John E. "The Bicentennial, the Philadelphia College of Pharmacy and Science, and early American pharmacy." AMER J PHARM 148(4) Jul/Aug '76, 101-7 (5 ref.)

-----. See also UNITED STATES PHARMACOPEIA 990

697 PHILADELPHIA, PENNSYLVANIA

"Philadelphia--three ages of a city." CIVIL ENG ASCE 46(7) Jul '76, 44-56

PHLOGISTON. See PRIESTLEY, JOSEPH 734-35; SCIENCE--18th CENTURY 803

698 PHOTOGRAMMETRY

Quinn, A. O. "Two centuries of service."

PHOTOGRAMMET ENG REMOTE SENS 42(7) Jul '76,
953-58 (12 ref.)

PHOTOGRAPHY. See DRAPER, JOHN WILLIAM 237

699 PHOTOGRAPHY--HIGH SPEED

Hyzer, William G. "Photoinstrumentation and
the SMPTE." SMPTE J 85(7) Jul '76, 547-51
(42 ref.)

700 PHRENOLOGY

"Phrenology comes to America." HOSP COMMUNITY
PSYCHIATRY 27(7) Jul '76, 484

701 -----

Tucker, Shelia. "Phrenology, bumps or bust."
NY STATE J MED 76(7) Jul '76, 1141-42

702 -----

Walsh, Anthony A. "Phrenology and the Boston
medical community in the 1830's." BULL HIST
MED 50(2) Sum '76, 261-73 (49 ref.)

PHYSIATRY. See PHYSICAL MEDICINE 703

703 PHYSICAL MEDICINE

Ruskin, Asa P. "Physiatry: physical medicine
and rehabilitation, past, present, and future." NY STATE J MED 76(8) Aug '76, 1355-59
(3 ref.)

704 PHYSICAL THERAPY

"The beginning of 'modern physiotherapy'."
PHYS THER 56(1) Jan '76, 15-21 (2 ref.)

705 -----

Carlin, Eleanor J. "The revolutionary spirit."
PHYS THER 56(10) Oct '76, 1111-16 (1 ref.)

706 -----

Granger, F. B. "The development of physiotherapy." PHYS THER 56(1) Jan '76, 13-14

707   -----

"Recollections and reminiscences from former reconstruction aides." PHYS THER 56(1) Jan '76, 22-40

PHYSICIANS. See MUNSON, ENEAS 630

708   PHYSICIANS--18th CENTURY

Miller, Genevieve. "A physician in 1776." AMER MED ASSN J 236(1) Jul 5, '76, 26-30 (15 ref.)

709   -----

"Signers of the Declaration of Independence, physician-patriots." MED ASSN STATE ALA J 45 (11) May '76, 22-23

710   -----

Spillane, J. D. "Doctors of 1776." BRIT MED J 1(6025) 26 Jun, '76, 1571-74; 2 (6026) 3 Jul, '76, 28-31 (30 ref.)

711   -----

Wallach, Gert. "The mind of the Colonial physician." CONN MED 40(12) Dec '76, 815-27 (123 ref.)

-----. See also WILSON, MATTHEW 1047

PHYSICIANS--19th CENTURY. See BRADLEY, SAMUEL BEACH 115

PHYSICIANS--ALABAMA. See HILL, LUTHER LEONIDAS, JR. 389

712   PHYSICIANS--DELAWARE

Shands, Alfred R., Jr. "Early Delaware doctors." DEL MED J 48(7) Jul '76, 379-89 (16 ref.)

PHYSICIANS--ILLINOIS. See CATTERTON, DYLER 137; NICKERSON, L. H. A. 650

PHYSICIANS--ILLINOIS--19th CENTURY. See JAYNE,
GERSHON 457; WOMEN PHYSICIANS--ILLINOIS--
19th CENTURY 1055

713 PHYSICIANS--LONG ISLAND
Sammis, Estelle P. "Early Long Island docs,
dedicated, compassionate." NY STATE J MED
76(7) Jul '76, 1163-64

PHYSICIANS--NEW JERSEY. See ELMER, JONATHAN
263; ENGLISH, THOMAS DUNN 284; FORT, GEORGE
FRANKLIN 326

PHYSICIANS--NEW JERSEY--18th CENTURY. See
MIDDLETON, PETER 596; NEWELL, WILLIAM
AUGUSTUS 648

PHYSICIANS--NEW YORK--19th CENTURY. See MC
CLELLAND, WILLIAM 517

PHYSICIANS--OHIO--19th CENTURY. See DUTTON,
CHARLES 241

714 PHYSICIANS--PENNSYLVANIA--18th CENTURY
Wenger, Donna F. "Tribute to Pennsylvania's
earliest physicians." PENN MED 79(7) Jul '76,
64-67

PHYSICIANS--RHODE ISLAND. See FULTON, FRANK
TAYLOR 349; PORTER, GEORGE WHIPPLE 729;
WALKER, MARY EDWARDS 1003

PHYSICISTS. See WOOD, ROBERT WILLIAMS 1056

715 PHYSICK, PHILIP SYNG
Marmelzat, Willard L. "Philip Syng Physick,
the reluctant medical student who became the
'father of American surgery', a saga for the
Bicentennial." J DERMATOL SURG 2(5) Nov '76,
380-81, 419-20 (5 ref.)

716 PHYSICS--PICTORIAL WORKS
"Two hundred years of American physics."

PHYS TODAY 29(7) Jul '76, 23-31

PHYSICS EDUCATION. See HALL, EDWIN 376;
RICHTMYER, FLOYD K. 771-72

717 PHYSICS, INDUSTRIAL

Weart, Spencer R. "The rise of 'prostituted'
physics." NATURE 262(5563) Jul 1, '76, 13-17
(35 ref.)

718 PIPE, CLAY

"Clay Pipe in America, a beginning on the
potter's wheel." BRICK CLAY REC 169(1) Jul
'76, 32

719 PLANT PROTEINS

Boyer, Robert A. "Early history of the plant
protein industry." CEREAL FOOD WORLD 21(7)
Jul '76, 294-95, 297-98

PLANTS. See BOTANY--NOMENCLATURE 111

720 PLUTONIUM

Gwynne, Peter. "Plutonium: 'free' fuel or in-
vitation to a catastrophe?" SMITHSONIAN 7(4)
Jul '76, 92-99

721 POLIOMYELITIS--THERAPY

"Infantile Paralysis, pioneers in treatment."
PHYS THER 56(1) Jan '76, 42-49

722 PONY EXPRESS

Herndon, Paul C. "The story of the Pony Express, ten days to San Francisco." OUR PUBLIC
LANDS 26(3) Sum '76, 12-19

723 -----

White, Joyce. "Pony Express riders and stage-
coach drivers." WEST HORSE 41(1) Jan '76,
22-23, 115-17; (2) Feb '76, 30-32, 100-102

724 PONY EXPRESS--UTAH

Headley, John W. "Survey for history." OUR

PUBLIC LANDS 26(3) Sum '76, 10-11

725 POPULATION GENETICS

Cross, Harold E. "Population studies and the old order Amish." NATURE 262(5563) Jul 1, '76, 17-20 (19 ref.)

726 POPULATION, INCREASE OF

Borgstrom, Georg. "The numbers force us into a world like none in history." SMITHSONIAN 7(4) Jul '76, 70-74, 76-77

727 -----

"Two centuries of population growth." METRO-POL LIFE STAT BULL 57, Jul-Aug '76, 2-6

728 PORTAIL, LOUIS LE BÈGUE

Buzzaird, Raleigh B. "Washington's most brilliant engineer." MIL ENG 68(442) Mar-Apr '76, 104-10

729 PORTER, GEORGE WHIPPLE

Goldowsky, Seebert J. "George Whipple Porter, M.D., 1847-1910." RHODE ISLAND MED J 59(2) Feb '76, 65-68, 74-76 (8 ref.)

730 PORTSMOUTH, NEW HAMPSHIRE

Estes, J. Worth. "'As healthy a place as any in America': Revolutionary Portsmouth, N.H." BULL HIST MED 50(4) Win '76, 536-52 (38 ref.)

731 PREGNANCY COMPLICATIONS

Greenwood, Ronald D. "Early descriptions in medicine: obstetrical complications." MD STATE MED J 25(5) May '76, 53-55 (3 ref.)

732 PRENATAL CARE--19th CENTURY

McGrellis, Nyra M. "Prenatal care 120 years ago." JOGN 5(2) Mar/Apr '76, 56-58

733   PRESCHOOL EDUCATION--SCIENCE
      McIntyre, Margaret. "A Bicentennial glance
      at preschool science." SCI & CHILD 13(4) Jan
      '76, 22-23 (7 ref.)
      PRESIDENTS--UNITED STATES. See LONGEVITY--
      UNITED STATES--FIRST LADIES 511; LONGEVITY--
      UNITED STATES--PRESIDENTS 512
      PRESIDENTS--UNITED STATES--VISION. See VISION--
      UNITED STATES--PRESIDENTS 1003
734   PRIESTLEY, JOSEPH
      Raman, V. V. "Joseph Priestley, an early im-
      migrant scientist." PHYS TEACH 14(6) Sep '76,
      335-39 (7 ref.)
735   -----
      Schofield, Robert E. "A discourse on the
      branches of natural philosophy most particu-
      larly related to chemistry." J CHEM ED 53(7)
      Jul '76, 409-13
      -----. See also WOODHOUSE, JAMES 1057
      PRINCETON, BATTLE OF 1777- WEATHER. See TREN-
      TON, BATTLE OF, 1776--WEATHER 914
736   PROHIBITION
      Rosa, Nicholas. "Rumrunning." OCEANS 9(1)
      Jan-Feb '76, 32-37
737   PSYCHIATRIC HOSPITALS
      "Asylum: a late 19th century view." HOSP
      COMMUNITY PSYCHIATRY 27(7) Jul '76, 485-89
738   -----
      "The Kirkbride plan." HOSP COMMUNITY PSY-
      CHIATRY 27(7) Jul '76, 473-77
739   PSYCHIATRY
      Braceland, Francis J. "A Bicentennial ad-
      dress: Benjamin Rush and those who came after

him." <u>AMER</u> <u>J</u> <u>PSYCHIAT</u> 133(11) Nov '76, 1251-
58 (22 ref.)

740 -----

"The mathematical curability of insan-
ity . . . and its attack by Pliny Earle."
<u>HOSP</u> <u>COMMUNITY</u> <u>PSYCHIATRY</u> 27(7) Jul '76, 481

741 -----

Menninger, R. W. "Psychiatry 1976: time for
a holistic medicine." <u>ANN</u> <u>INTERN</u> <u>MED</u> 84(5)
May '76, 603-4

742 -----

"Moral treatment in America's lunatic asy-
lums." <u>HOSP</u> <u>COMMUNITY</u> <u>PSYCHIATRY</u> 27(7) Jul
'76, 468-70

743 -----

Ozarin, Lucy D., Richard W. Redick, and Carl
A. Taube. "A quarter century of psychiatric
care, 1950-1974: a statistical review." <u>HOSP</u>
<u>COMMUNITY</u> <u>PSYCHIATRY</u> 27(7) Jul '76, 515-19
(3 ref.)

744 PSYCHIATRY--NEW YORK
Mora, George. "Psychiatry in New York State
from early times to end of nineteenth cen-
tury." <u>NY</u> <u>STATE</u> <u>J</u> <u>MED</u> 76(7) Jul '76, 1201-04

745 PSYCHIATRY--PICTORIAL WORKS
"200 years of American psychiatry." <u>MED</u> <u>WORLD</u>
<u>N</u> 17(23) Oct 25, '76, 68-73

746 PSYCHOLOGY
McKeachie, Wilbert J. "Psychology in Ameri-
ca's Bicentennial year." <u>AMER</u> <u>PSYCHOL</u> 31(12)
Dec '76, 819-33 (75 ref.)

747 PUBLIC HEALTH--INTERNATIONAL
Knowles, J. H. "American medicine and world

health 1976." <u>ANN</u> <u>INTERN</u> <u>MED</u> 84(4) Apr '76, 483-85 (18 ref.)

748 PUBLIC HEALTH--NEW YORK CITY--18th CENTURY
Duffy, John. "Public health in New York City in the Revolutionary period." <u>AMER</u> <u>MED</u> <u>ASSN</u> <u>J</u> 236(1) Jul 5, '76, 47-51 (26 ref.)

749 PUBLIC HEALTH--POLICY
Steinfeld, Jesse L. "National health policy 1976." <u>ANN</u> <u>INTERN</u> <u>MED</u> 85(5) Nov '76, 669-71

750 PUBLIC LANDS
Nelson, A. Z. "Our original public domain." <u>AMER</u> <u>FOR</u> 82(2) Feb '76, 28-31, 66-67; (3) Mar '76, 46-49 (33 ref.)

751 PUBLIC LANDS--LAWS AND REGULATIONS
Herndon, Paul C. "Establishing a policy for the public lands." <u>OUR</u> <u>PUBLIC</u> <u>LANDS</u> 26(3) Sum '76, 20-22

752 PUBLIC LANDS--OKLAHOMA
Rice, Anthony F. "Land rush! The opening of Oklahoma's Cherokee Outlet from the vantage point of an eye witness." <u>OUR</u> <u>PUBLIC</u> <u>LANDS</u> 26(3) Sum '76, 4-9

753 PUBLISHERS AND PUBLISHING
Kuney, Joseph H. "Scientific and technical publishing-1776-1976." <u>AMER</u> <u>SOC</u> <u>INFORM</u> <u>SCI</u> <u>BULL</u> 2(8) Mar '76, 30-31

754 PUTNAM, RUFUS
Buzzaird, Raleigh B. "Washington's favorite engineer." <u>MIL</u> <u>ENG</u> 68(444) Jul-Aug '76, 298-301

PUZZLES. See LOYD, SAM 514

PYTHAGOREAN THEOREM. See Garfield, James 355

755 QUALITY CONTROL
       Golomski, William A. "Quality control-his-
       tory in the making." QUAL PROG 9(7) Jul '76,
       16-18
756 -----
       McCreary, Robert M. "Whence cometh quality
       control?" QUAL PROG 9(7) Jul '76, 15(7 ref.)
757 QUILTING
       Madden, Betty. "200 years of American quilts
       in Illinois collections." LIVING MUS 38(6)
       Nov-Dec '76, 418-20
       QUININE. See FEVER 297
       QUINSEY. See WASHINGTON, GEORGE--LAST ILLNESS
       AND DEATH 1015
758 RAILROADS
       Morgan, David P. "America, the land that
       railroads made possible." TRAINS 36(9) Jul
       '76, 20-51
759 -----
       Stover, John F. "Railroads and the building
       of America." RAILWAY AGE 177(12) Jul 4, '76,
       52-63
760 -----
       White, John H., Jr. "American railroads: a
       Bicentennial overview." RAILWAY AGE 177(12)
       Jul 4, '76, 64-65
       -----. See also ELECTRIC RAILROADS 252; PANAMA
       RAILROAD 681; STEAMSHIPS--19th CENTURY 840
761 RAILROADS--LAWS AND REGULATIONS
       "A Bicentennial Bill of Rights for America's
       railroads." RAILWAY AGE 177(12) Jul 4, '76,
       27-37

762 RAILWAY AGE (PERIODICAL)
     "Railway age: the first 100 years." RAILWAY
     AGE 177(12) Jul 4, '76, 100-102
763 RAPID TRANSIT
     Middleton, William D. "Rail transit and the
     building of the American city." RAILWAY AGE
     177(12) Jul 4, '76, 68-75
764 RAY, ISAAC
     "Ray's enduring legal treatise." HOSP COM-
     MUNITY PSYCHIATRY 27(7) Jul '76, 480
     REDFIELD, WILLIAM C. See WEATHER 1033
765 REFRACTORY MATERIALS
     "Refractories growth triggered by an indus-
     trial revolution." BRICK CLAY REC 169(1) Jul
     '76, 28-29
766 REFRIGERATION AND REFRIGERATING MACHINERY
     Nagengast, Bernard A. "The revolution in
     small vapor compression refrigeration."
     ASHRAE J 18(7) Jul '76, 36-40
767 REFRIGERATORS
     "The refrigerator: an American institution."
     EXXON CHEM MAG 9(3) '76, 16-17
     RESEARCH, INDUSTRIAL. See AMERICAN INSTITUTE
     OF ELECTRICAL ENGINEERS 31; EDISON, THOMAS
     A. 247; PHYSICS, INDUSTRIAL 717; STEINMETZ,
     CHARLES P. 844; TECHNOLGY--CHRONOLOGY 868
768 RESTAURANTS
     Evans, Evelyn. "200 years of great American
     inns." FOOD SERV MKTG 38(7) Jul '76, 29-31,
     33, 37

RESTAURANTS. See also FOOD INDUSTRY--19th CEN-
TURY 314

769 REVERE, PAUL
Hirschfeld, Fritz. "Paul Revere rides again."
MECH ENG 98(7) Jul '76, 18-19

770 RHODE ISLAND--MEDICAL ASPECTS
Farrell, John E. "The air of Rhode Island is
good." RHODE ISLAND MED J 59(7) Jul '76,
317-27, 339-42 (22 ref.)

771 RICHTMYER, FLOYD KARKER
Andrews, C. L. "Memories of a great teacher."
PHYS TEACH 14(1) Jan '76, 27-29 (2 ref.)

772 -----
Miner, Thomas D. "Floyd Karker Richtmyer."
PHYS TEACH 14(1) Jan '76, 26

773 RICKMAN, WILLIAM
Warthen, Harry J. "Doctor Rickman and Virginia's continental surgeons." VIR MED MON
103(7) Jul '76, 499-504 (24 ref.)

774 RITTENHOUSE, DAVID
"David Rittenhouse--American scientist and
innovator." DESIGN N 31(13) Jul 4, '76, 16

775 -----
"David Rittenhouse (1732-1796)." ASTRONOMY
4(7) Jul '76, 38
-----. See also WISTAR, CASPAR 1052

776 ROWLAND, HENRY
Miller, John D. "Rowland's physics." PHYS
TODAY 29(7) Jul '76, 39-45 (10 ref.)
RUHMKORFF, HEINRICH D. See PAGE, CHARLES G.
679

777 RUSH, BENJAMIN
Adcock, Louis H. "Benjamin Rush, America's

first professor of chemistry." CHEMISTRY 48
(11) Dec '75, 10-11, 20 (6 ref.)

778 -----

Carlson, Eric T., and Jeffrey L. Wollock.
"Benjamin Rush on politics and human nature."
AMER MED ASSN J 236(1) Jul 5, '76, 73-77 (20
ref.)

779 -----

Jones, Robert E. "Franklin and Rush." HOSP
COMMUNITY PSYCHIATRY 27(7) Jul '76, 461-63

780 -----

Rosen, George. "Benjamin Rush on health and
the American Revolution." AMER J PUB HEALTH
66(4) Apr '76, 397-98 (7 ref.)

781 -----

Shimkin, Michael B. "Benjamin Rush on cancer." CANCER RES 36(7 pt. 1) Jul '76, 2117-
18 (2 ref.)

782 -----

Veith, Ilza. "Benjamin Rush and the beginnings of American medicine." WEST J MED 125
(1) Jul '76, 17-27 (6 ref.)
-----. See also ALCOHOLISM--18th CENTURY 20;
MEDICINE--18th CENTURY 567; PSYCHIATRY 739

783 RUSSELL, HENRY NORRIS
"Henry Norris Russell (1877-1957)." ASTRON-
OMY 4(7) Jul '76, 93

784 RUTHERFURD, LEWIS MORRIS
Devons, Samuel. "Lewis Morris Rutherfurd,
1816-1892." APP OPTICS 15(7) Jul '76, 1731-
40 (21 ref.)

785 SAFETY ENGINEERING
     "A Bicentennial look at safety." PROF SAFETY
     21(4) Apr '76, 16-25
786 SAILING VESSELS
     "Storybook ships and sail races." NOAA 6(4)
     Oct '76, 20-21
787 -----
     "The tall ships, spectular stars of OpSail
     76." MARINE ENG/LOG 31(8) Jul '76, 27-34
788 -----
     Weigel, Edwin P. "When the tall ships came."
     NOAA 6(4) Oct '76, 12-19
789 ST. JOHN'S HOSPITAL (SPRINGFIELD, ILLINOIS)
     Rolens, M. E. "One hospital--100 years."
     ILL MED J 149(2) Feb '76, 161-62
790 ST. VINCENT'S HOSPITAL (NEW YORK CITY)
     Ollstein, Ronald N., Edward C. Haggerty, and
     Peter Rosenthal. "St. Vincent's Hospital and
     Medical Center of New York, one hundred
     twenty-five years." NY STATE J MED 76(2) Feb
     '76, 306-9 (2 ref.)
     SALVAGE. See BON HOMME RICHARD 107
791 SANITARY CONSULTING ENGINEERING
     White, Robert L. "Opportunity knocks in de-
     veloping countries." WATER WASTES ENG 13(7)
     Jul '76, 77-80
792 SANITARY CONSULTING ENGINEERING--FORECASTS
     Culver, Robert H., A. A. Kalinske, and
     Richard L. Woodward. "Innovation is an old
     idea-with a big future." WATER WASTES ENG
     13(7) Jul '76, 42-44, 46, 48, 112
793 -----
     Heckroth, Charles W. "Will the real consult-

ing sanitary engineer please stand up."
WATER WASTES ENG 13(7) Jul '76, 39-41, 102
SANITARY ENGINEERING. See WARING, GEORGE 1007
794 SANITARY ENGINEERING--FORECASTS
Storck, William J. "How big will CE's get?"
WATER WASTES ENG 13(7) Jul '76, 35-37
795 SANITATION
Bloch, Harry. "Sanitary awakening in America." MED SOC NJ J 73(5) May '76, 435-37 (27 ref.)
796 SANITATION--SURVEYS
Rosen, George. "John Shaw Billings and the plan for a sanitary survey of the United States." AMER J PUB HEALTH 66(5) May '76, 492-95 (16 ref.)
SAUGUS IRON WORKS. See STEEL INDUSTRY--18th CENTURY 842
797 SCABIES
"The Itch." MORBID MORTAL WEEK REP 25(25 pt. 2) Jul 2, '76, 3 (1 ref.)
SCARLATINA. See STREPTOCOCCAL INFECTIONS--19th CENTURY 847
798 SCHOOLS, MEDICAL--NEW YORK
"Medical schools." NY STATE J MED 76(7) Jul '76, 1205, 1208-09, 1213-18.
799 SCIENCE
Reingold, Nathan. "Reflections on 200 years of science in the United States." NATURE 262 (5563) Jul 1, '76, 9-13
800 -----
Richards, Graham. "America's contribution to science." NEW SCI 72(1031) Dec 16, '76, 642-44

801 -----
     Wakeling, Patricia. "Bicentennial." APP OP-
     TICS 15(7) Jul '76, 1695-1705
802 SCIENCE--18th CENTURY
     Caccia, Bernard. "Science in Revolutionary
     America." GOLD BULL 9(3) Jul '76, 91-103
803 -----
     Harken, Alden H. "Oxygen, politics and the
     American Revolution (with a note on the Bi-
     centennial of phlogiston)." ANN SURG 184(5)
     Nov '76, 645-50 (21 ref.)
804 -----
     Thackray, Arnold. "Scientific networks in
     the age of the American Revolution." NATURE
     262(5563) Jul 1, '76, 20-24
     SCIENCE--GREAT BRITAIN. See SCIENCE--INTER-
     NATIONAL ASPECTS 805
805 SCIENCE--INTERNATIONAL ASPECTS
     Hodgkin, Alan. "The early years of Anglo-
     American co-operation in science." INTER-
     DISCIP SCI R 1(2) Jun '76, 109-19 (25 ref.)
806 SCIENCE--SPANISH AMERICA
     Roche, Marcel. "Early history of science in
     Spanish America." SCIENCE 194(4267) Nov 19,
     '76, 806-10 (46 ref.)
807 SCIENCE--STUDY AND TEACHING--GEORGIA
     Stanitski, Conrad L. "Georgia's science
     teachers and science teaching, an historical
     description." GA ACAD SCI BULL 34(4) Sep '76,
     216-23 (13 ref.)
     SCIENCE--TENNESSEE. See BROWN, LUCIUS POLK 122
808 SCIENTISTS
     Greenberg, Daniel S. "Scientists wanted--

pioneers needn't apply; call AD 2000." SMITH-
SONIAN 7(4) Jul '76, 60-64, 66-67

809 SCULPTURE
Forgey, Benjamin. "Two centuries of U.S.
sculpture all in one place." SMITHSONIAN 7
(6) Sep '76, 52-59

810 SEA WATER--DESALTING--18th CENTURY
Hirschfeld, Fritz. "Desalination-circa 1790."
MECH ENG 98(6) Jun '76, 20-21

811 SEAMEN--19th CENTURY
Dye, Ira, "Early American merchant seafar-
ers." AMER PHIL SOC PROC 120(5) Oct 15, '76,
331-60 (48 ref.)

812 SEWAGE DISPOSAL
Rowland, W. G., Jr. and A. S. Heid. "Water
and the growth of the nation." WATER POLLUT
CONTROL FED J 48(7) Jul '76, 1682-89 (10
ref.)

813 SEWAGE, UTILIZATION OF--19th CENTURY
Tarr, Joel A. "City wastes in the 1800's
made it back to the farm." ORGANIC GARD FARM
23(12) Dec '76, 74-77

814 SHAPLEY, HARLOW
"Harlow Shapley (1885-1972)." ASTRONOMY
4(7) Jul '76, 94

815 SHEEP
Noh, Laird. "The sheep industry: its past
and future." NAT WOOL GROW 66(11) Nov '76,
18-19, 25-28

816 -----
Noh, Laird, "200 years in the sheep industry . . . and beyond." SHEEP BREED SHEEPMAN
96(7) Jul '76, 22, 24, 26, 28-30

817   -----
       "Sheep figured in Revolution as British
       curbed industry." NAT WOOL GROW 66(7) Jul '76,
       12
       -----. See also BELL, ALEXANDER GRAHAM 97;
       JEFFERSON, THOMAS 463, 467; MERINO SHEEP 592;
       WASHINGTON, GEORGE 1012
       SHEET COPPER. See REVERE, PAUL 769
818   SHIPBUILDING
       "U.S. merchant shipbuilding, 1607-1976."
       MARINE ENG/LOG 31(9) Aug '76, 65-77, 172
       (6 ref.)
       SHIPPEN, WILLIAM J. See MEDICINE--18th CENTURY
       567
819   SHIPS IN ART
       Petrides, Bette. "Revival of marine art in
       America." OCEANS 9(5) Sep-Oct '76, 8-13
820   SHIPYARDS--18th CENTURY
       Soderholm, Lars. "Yankee shipyards prosper."
       DESIGN N 31(13) Jul 4, '76, 24-25
       SILLIMAN, BENJAMIN. See SILLIMAN, BENJAMIN, JR.
       821
821   SILLIMAN, BENJAMIN, JR.
       Scott, Arthur F. "Notes from the 1874 essay
       of Benjamin Silliman, Jr." CHEMISTRY 49(6)
       Jul-Aug '76, 8-11
       SILVER MINES AND MINING. See MINES AND MINERAL
       RESOURCES--NEVADA 605
822   SIMS, J. MARION
       Sparkman, Robert S. "A woman's surgeon, Dr.
       J. Marion Sims." MED ASSN STATE ALA J 45(11)
       May '76, 19-22

SISTER MARY JOSEPH. See UMBILICAL CANCER 917

823 SKELETON--MAN

Angel, J. Lawrence. "Colonial to modern
skeletal change in the U.S.A." AMER J PHYS
ANTHROP 45(3, pt. 2) Nov '76, 723-35 (56
ref.)

824 SKYSCRAPERS

Dallaire, Gene. "Birth of the skyscraper."
CIVIL ENG-ASCE 46(7) Jul '76, 61-64 (2 ref.)

SLATER, SAMUEL. See TEXTILE INDUSTRY 894

825 SLAVE TRADE--PICTORIAL WORKS

Howell, Keith K. "The infamous Blackbirders."
OCEANS 9(1) Jan-Feb '76, 17-23

826 SLUDGE

Haines, Roger F. "Sludge-where will we put
it?" WATER WASTES ENG 13(7) Jul '76, 60, 62,
64, 66

SMALE, STEPHEN. See DYNAMICAL SYSTEMS 242

827 SMALL ANIMALS--VETERINARY MEDICINE

Drenan, David M. "The growth and development
of small animal practice in the United
States." AMER VET MED ASSN J 169(1) Jul 1,
'76, 42-49 (9 ref.)

828 SMALLPOX

"Inoculation against smallpokkes (variola-
tion)." MORBID MORTAL WEEK REP 25(25, pt. 2)
Jul 2, '76, 1-2

829 SMALLPOX--MILWAUKEE--19th CENTURY

Leavitt, Judith W. "Politics and public
health: smallpox in Milwaukee, 1894-1895."
BULL HIST MED 50(4) Win '76, 553-68 (44 ref.)

SMITH, ADAM. See WATT, JAMES 1031

SMITH, STEVEN. See EMIGRATION AND IMMIGRA-
TION--HEALTH   268

830   SOCIETIES, MEDICAL--19th CENTURY

Kaufman, Martin. "The admission of women to
nineteenth-century American medical societies." BULL HIST MED 50(2) Sum '76, 251-60
(35 ref.)

831   -----

Pavey, Charles W. "Doctors' dining clubs,
the beginning of continuing medical education." OHIO STATE MED J 72(1) Jan '76, 10-12

832   SOCIETIES, MEDICAL--ILLINOIS--19th CENTURY

Camp, Harold M. "Early medical societies."
ILL MED J 150(1) Jul '76, 49-54

833   SOCIETY OF MOTION PICTURE AND TELEVISION ENGI-
NEERS--BIOGRAPHY

Coleman, Leonard. "Honor role of the Society."
SMPTE J 85(7) Jul '76, 552-61

834   SOCIETY OF MOTION PICTURE AND TELEVISION EN-
GINEERS--EDUCATION

Farmer, Herbert E. "Education and the soci-
ety-1916-1930 as reported in the Transactions." SMPTE J 85(7) Jul '76, 561-68, 570

835   SOIL MECHANICS

Tschebotarioff, Gregory P. "Half a century
of soil mechanics--some thoughts for the
future in the light of the past." ENG ISSUES
(PROC. ASCE) 102(EI3) Jul '76, 321-37 (7 ref.)

SOLERI, PAOLO. See CITIES AND TOWNS 174

836   SPECTRUM--INFRARED

Barr, E. Scott. "Men and milestones in optics. VI: the rise of infrared spectroscopy
in the U.S.A. to World War II." APP OPTICS

15(7) Jul '76, 1707-21 (114 ref.)

SPECTRUM ANALYSIS--19th CENTURY. See ALTER, DAVID 23

-----. See RUTHERFURD, LEWIS MORRIS 784

SPECTRUM ANALYSIS--TABLES. See DIFFRACTION GRATINGS 232

SPEEDWELL SOCIETY. See INFANTS--MORTALITY 429

837 SPOTSWOOD, WILLIAM A. W.

Chitwood, W. R. "Doctor Spotswood and the Confederate Navy." VIR MED MON 103(10) Oct '76, 729-33 (30 ref.)

STARS, VARIABLE. See LEAVITT, HENRIETTA 480

838 STATE TREES

"Bicentennial history." AMER FOR 82(3) Mar '76, 40; (4) Apr '76, 36; (5) May '76, 48; (6) Jun '76, 52; (7) Jul '76, 4; (8) Aug '76, 42; (9) Sep '76, 36; (10) Oct '76, 28; (11) Nov '76, 36; (12) Dec '76, 52

839 STATUE OF LIBERTY

Carlinsky, Dan. "Visit to Miss Liberty." MOD MATURITY 19(3) Jun-Jul '76, 43-45

840 STEAMSHIPS--19th CENTURY

Hirschfeld, Fritz. "Linking the nation by water and rail." MECH ENG 98(9) Sep '76, 20-23

841 STEEL INDUSTRY

Austin, James B. "200 years of U.S. iron and steel industry." QUAL PROG 9(7) Jul '76, 22-24

842 STEEL INDUSTRY--18th CENTURY

Sevrens, Palmer E. "Bog iron: Massachusetts mineral heritage." ROCK MIN 51(5) Jun '76

228-33 (3 ref.)

843 STEEL INDUSTRY--PICTORIAL WORKS
"A look at early steelmaking in America."
IRON & STEEL ENG 53(7) Jul '76, 62-67

844 STEINMETZ, CHARLES P.
Brittain, James E. "C.P. Steinmetz and E.F.W.
Alexanderson: creative engineering in a cor-
porate setting." IEEE PROC 64(9) Sep '76,
1413-17 (39 ref.)

845 STORMS
Hayden, Everett, "The great storm off the
Atlantic coast of the United States, March
11-14, 1888." MARINE TECH SOC J 10(6) Jul-
Aug '76, 29-33

846 STOVES
"Cooking appliances come of age: from stones
to microwaves." EXXON CHEM MAG 9(3) '76, 18-
19

847 STREPTOCOCCAL INFECTIONS--19th CENTURY
Greenwood, Ronald D. "An account of scarlatina epidemic, 1839." ILL MED J 150(2) Aug
'76, 147-48 (2 ref.)

848 STRUCTURAL ENGINEERING
Randall, Frank A., Jr. "The safety factor of
structures in history." PROF SAFETY 21(1)
Jan '76, 12-18 (12 ref.)

849 STUDY COMMITTEE OF THE WISCONSIN MATERNAL MOR-
TALITY SURVEY
Leonard, Thomas A. "History of the Study
Committee of the Wisconsin Maternal Mortality
Survey." WIS MED J 75(9) Sep '76, 29-32

850 SUBMARINE-WARFARE--18th CENTURY
Hirschfeld, Fritz. "Submarine warfare-

General Washington's secret weapon." <u>MECH</u>
<u>ENG</u> 98(12) Dec '76, 18-19

851 -----

Smoluk, George R. "British attacked by sub-
marines and mines." <u>DESIGN</u> <u>N</u> 31(13) Jul 4,
'76, 18

SULLIVAN, JOHN, GENERAL. SEE LONG ISLAND, BAT-
TLE OF, 1776--MEDICAL ASPECTS 507

SURGEONS. See HILL, LUTHER LEONIDAS, JR. 389;
JONES, JOHN 472; PHYSICK, PHILIP SYNG 715

SURGEONS--19th CENTURY. See TAYLOR, CHARLES
FAYETTE 861

852 SURGEONS--CONNECTICUT

Roberts, Melville. "Connecticut's first sur-
geon: Thomas Pell." <u>CONN</u> <u>MED</u> 40(12) Dec '76,
856-57 (9 ref.)

SURGEONS--ILLINOIS. See BYRD, WILLIAM A. 127

SURGEONS--KANSAS. See HERTZLER, ARTHUR 387

SURGEONS--VIRGINIA--18th CENTURY. See RICKMAN,
WILLIAM 773

SURGERY--CALGARY, CANADA. See MEDICINE--CAL-
GARY, CANADA 570

853 SURGERY--RHODE ISLAND

Perry, Thomas, Jr. "Surgery in Rhode Island
after the American Revolution." <u>RHODE</u> <u>ISLAND</u>
<u>MED</u> <u>J</u> 59(3) Mar '76, 105-8, 138 (11 ref.)

854 SURGERY--WISCONSIN

Falk, Victor S. "Wisconsin's surgical heri-
tage." <u>WIS</u> <u>MED</u> <u>J</u> 75(4) Apr '76, 24-28 (9
ref.)

SURGERY, GYNECOLOGIC. See SIMS, J. MARION 822

855 SURGERY, PLASTIC--CONNECTICUT--18th CENTURY

Arons, Marvin S., M. Felix Freshwater, and

Richard Hegel. "Abel's Auricle-a colonial
tale of plastic surgery." CONN MED 40(12)
Dec '76, 851-55 (27 ref.)

SURVEYING. See DE BRAHM, WILLIAM GERARD 220;
ELLICOTT, ANDREW 262; GEODESY 358; GUNTER,
EDMUND 373

856 SURVEYING--18th CENTURY
Steele, Poly A. "Charting the Colonies: the
art of surveying and instrument-making."
DESIGN N 31(13) Jul 4, '76, 20-21

857 SURVEYING--FORECASTS
Wolf, Paul R. "Surveying: current status and
future challenges." SURVEY MAP 36(2) Jun '76,
155-60

858 SWINE
Black, Roe, and John Russell. "Hogs take
first prize in bicen competition." FARM J
100(3) mid-Feb '76, J-1-J-3

859 SYCAMORE SHOALS, TENNESSEE--18th CENTURY
Fielder, George F., and David Higgs. "Syca-
more Shoals: typical of frontier Tennessee."
TENN CONSERV 42(6) Jul '76, 12-13

860 SYPHILIS
"The Large Pokkes--July, 1776." MORBID MOR-
TAL WEEK REP 25(25 pt. 2) Jul 2, '76, 5 (4
ref.)

861 TAYLOR, CHARLES FAYETTE
Shands, Alfred R. "Charles Fayette Taylor
and his times--1827 to 1899." SURG GYNECOL
OBSTET 143(5) Nov '76, 811-18 (28 ref.)

862 TEA--THERAPEUTIC USE--18th CENTURY
Greden, John F. "The tea controversy in
Colonial America." AMER MED ASSN J 236(1)

Jul 5, '76, 63-66 (15 ref.)

863 TECHNOLOGICAL CHANGE

Kane, J.T. "Technology and limits of growth: the great American dialogue after two centuries of expansion." PROF ENG 46(1) Jan '76, 14-26

864 TECHNOLOGICAL CHANGE--SOCIAL ASPECTS

Gilpin, Robert. "Exporting the technological revolution." SAT REV 3(6) Dec 13, '75, 31-32, 35-36

865 TECHNOLOGY

Clark, Wilson, "Big and/or little? Search is on for right technology." SMITHSONIAN 7(4) Jul '76, 42-49

866 -----

Dibner, Bern. "Bern Dibner of Burndy speaks on our technological heritage." ELECTRONIC DESIGN 24(4) Feb 16, '76, 130-33

867 -----

Jones, Robert R. "Technology's triumphs." IND RES 18(12) Nov 15, '76, 10, 12, 14, 16

868 TECHNOLOGY--CHRONOLOGY

"Industrial Research magazine's Bicentennial chronology of American technology." IND RES 18(12) Nov 15, '76, 41-42, 44, 46, 48-51

-----. See also MINERALS 601

869 TELECOMMUNICATION

Bell System Centennial: 100 years of publishing on telecommunications." BELL SYSTEM TECH J 55(3) Mar '76, 273-76

870 -----

Everitt, William L. "Telecommunications--the resource not depleted by use. A historical

and philosophical resumé." IEEE PROC 64(9)
Sep '76, 1292-99

871 -----
Grossman, Morris. "The communications era
(1879-1905) extending man's voice by wire
and radio." ELECTRONIC DESIGN 24(4) Feb 16,
'76, 88-95

872 -----
Kaye, David N. "The vacuum tube era (1905-
1948) taking the crucial step for modern
technology." ELECTRONIC DESIGN 24(4) Feb 16,
'76, 98-105

873 -----
Nilles, Jack M. "Talk is cheaper." IEEE
SPECTRUM 13(7) Jul '76, 90-93

874 TELECOMMUNICATION--FORECASTS
Barnouw, Erik. "So you think TV is hot
stuff? Just you wait." SMITHSONIAN 7(4) Jul
'76, 78-84

875 TELEPHONE
Altepeter, Henry M., "The birth of our telephone industry." TELEPHONY 190(23) Jun 7,
'76, 50-51

876 -----
Altepeter, Henry M. "The early days of
telephony." TELEPHONY 190(1) Jan 5, '76,
60-62

877 -----
Altepeter, Henry M. "From the 'wireless' to
'Telstar' and beyond." TELEPHONY 190(21) May
24, '76, 82, 84, 88, 104

878 -----
Altepeter, Henry M. "The people and

principles that made telephony." TELEPHONY
191(14) Oct 4, '76, 124, 126, 128-31

879 -----

Bell, A. Graham. "Researches in telephony."
BELL SYSTEM TECH J 55(3) Mar '76, 279-88

880 -----

Maddox, Brenda. "A woman's place is at the
switchboard." NEW SCI 69(992) Mar 18, '76,
614-15

881 -----

Watson, T.A. "The birth and babyhood of the
telephone: Dr. Watson's address (1928)."
POST OFF ELECTR ENG J 69(1) Apr '76, 3-11

882 -----

Wolff, Michael F. "The marriage that almost
was." IEEE SPECTRUM 13(2) Feb '76, 40-51
-----. See also BELL, ALEXANDER GRAHAM 96-101

883 TELEPHONE--19th CENTURY
Bocock, John P. "Will the real inventor of
the telephone please stand up." TELEPHONY
191(21) Nov 22, '76, 76-79

884 TELEPHONE--CHRONOLOGY
"Telephone calendar." TELEPHONY 191(1) Jul
5, '76, 226, 232, 234-40, 244-45, 251-59,
261-62

885 TELEPHONE--DENVER--19th CENTURY
Hardin, Helen. "Telephony in the wild, wild
west." TELEPHONY 191(1) Jul 5, '76, 182, 184,
188, 190, 192

886 TELEPHONE--FORECASTS
Campbell, Duncan. "The future in store for
the telephone." NEW SCI 69(992) Mar 18, '76,
612-14

887 TELEPHONE--ILLINOIS
     Keelyn, J.E. "The story of Illinois." TELE-
     PHONY 190(4) Jan 26, '76, 40-41
888 TELEPHONE--OHIO
     Critchfield, H.D. "The telephone story of
     Ohio." TELEPHONY 190(21) May 24, '76, 80-81
889 TELEVISION
     Fink, Donald G. "Perspectives on television:
     the role played by the two NTSC's in pre-
     paring television service for the American
     public." IEEE PROC 64(9) Sep '76, 1322-31
     (20 ref.)
890 -----
     O'Brien, Richard S., and Robert B. Monroe.
     "101 years of television technology." SMPTE
     J 85(7) Jul '76, 457-80 (31 ref.)
     -----. See also VIOLENCE 1001
     TELEVISION LIGHTING. See MOTION PICTURE
     STUDIOS--LIGHTING 619
     TELEVISION SOUND RECORDING--MAGNETIC RECORD-
     ING. See MOTION PICTURE SOUND RECORDING--
     MAGNETIC RECORDING 618
891 TELEVISION, COLOR
     Herold, Edward W. "A history of color tele-
     vision displays." IEEE PROC 64(9) Sep '76,
     1331-38 (63 ref.)
     TEMPERANCE. See ALCOHOLISM 20-21
892 TEMPERANCE--MEDICAL ASPECTS--19th CENTURY
     Cassedy, James H. "An early American hang-
     over: the medical profession and intemper-
     ance, 1800-1860." BULL HIST MED 50(3) Fal
     '76, 405-13 (20 ref.)

893 TENNESSEE VALLEY AUTHORITY
     McCraw, Thomas K. "Triumph and irony--the
     TVA." IEEE PROC 64(9) Sep '76, 1372-80 (17
     ref.)
894 TEXTILE INDUSTRY
     Herschfeld, Fritz. "Samuel Slater meets
     Moses Brown and a textile empire is founded."
     MECH ENG 98(11) Nov '76, 16-17
895 -----
     "Textiles salutes the spirit of '76."
     TEXTILE WORLD 126(1) Jan '76, 57-65, 67
896 -----
     "200 years of U.S.A. textile manufacturing."
     AMER TEXTILE REP AT5(3) Mar '76, 29-34, 37-38,
     40, 42, 44, 46, 48, 52, 54, 58, 61-62, 64-
     66, 68, 70-72, 74, 76-82, 84-88
897 TEXTILE INDUSTRY--18th CENTURY
     Norwick, Braham. "1776, the textile industry
     and quality." QUAL PROG 9(7) Jul '76, 19-21
898 TEXTILE INDUSTRY--CHRONOLOGY
     Carter, C.W. "U.S. textile chronology." QUAL
     PROG 9(7) Jul '76, 20
899 TEXTILES
     "Bicentennial of American textiles, 1776-
     1976." AMER FABRIC FASHION (106) Win/Spr '76,
     61-68; (107) Spr '76, 44-58
     THERAPEUTICS. See DRUGS 240
900 THERAPEUTICS--18th CENTURY
     Parascandola, John. "Drug therapy in Coloni-
     al and Revolutionary America." AMER J HOSP
     PHARM 33(8) Aug '76, 807-10 (25 ref.)
901 THERAPEUTICS--19th CENTURY
     Greenwood, Ronald D. "A view of nineteenth

century therapeutics." MED ASSN STATE ALA J 45(12) Jun '76, 25

902 THOMPSON, BENJAMIN
Brown, Sanborn C. "Benjamin Thompson, Count Rumford." PHYS TEACH 14(5) May '76, 270-81 (38 ref.)

903 TILTON, JAMES
Saffron, Morris H. "The Tilton Affair." AMER MED ASSN J 236(1) Jul 5, '76, 67-72

904 TRADE ASSOCIATION--AUTOMOTIVE INDUSTRY
"The clay that binds." MOTOR AGE 95(6) Jun '76, 41-42, 45-46, 49-50, 53-54, 57-58

905 TRAFFIC ENGINEERING
Mueller, Edward A. "The transportation profession in the Bicentennial year." TRAFFIC ENG 46(7) Jul '76, 17-21; (9) Sep '76, 29-34; (11) Nov '76, 50-53

906 TRANSISTORS
Bursky, Dave. "The transistor era (1948-1959) approaching the age of space exploration." ELECTRONIC DESIGN 24(4) Feb 16, '76, 108-15
TRANSITS--COLONIAL PERIOD. See COMETS, COLONIAL PERIOD 190

907 TRANSPORTATION
Dallaire, Gene. "The story of America's transportation revolution." CIVIL ENG-ASCE 46(7) Jul '76, 70-76 (3 ref.)

908 -----
Mennie, Don. "People movers." IEEE SPECTRUM 13(7) Jul '76, 84-89

909 -----
Rawson, Bart, and Rich Cross. "Two hundred years of transportation." COMMERCIAL CAR J

130(6) Feb '76, 89-120

-----. See also TRAFFIC ENGINEERING 905

910 TRANSPORTATION--OHIO

Weed, J. Merrill. "A special Bicentennial
story for the people of Ohio featuring transportation." N ENG 48(3) Sum '76, 10-13, 21-
23

911 TREES

Lewis, Clarence E. "The American heritage
of trees." AMER NURSERYMAN 144(10) Nov 15,
'76, 14, 28, 30, 32, 34

912 -----

Lewis, Clarence E. "Trees in American history." AMER FOR 82(3) Mar '76, 50-54 (9 ref.)

-----. See also FLAGS 305; MARSHALL, HUMPHRY
530; STATE TREES 838; WASHINGTON, GEORGE
1011

913 TRENTON, BATTLE OF, 1776

Fleming, Thomas. "Christmas 1776, the way it
was." NAT WILDLIFE 15(1) Dec-Jan '77, 4-11

914 TRENTON, BATTLE OF, 1776--WEATHER

Ludlum, David M. "The weather of indepen-
dence--5: Trenton and Princeton." WEATHER-
WISE 29(2) Apr '76, 74-83 (9 ref.)

915 TUBERCULOSIS

"Consumption--Salem, Massachusetts." MORBID
MORTAL WEEK REP 25(25, pt. 2) Jul 2, '76, 3
(2 ref.)

916 TUCKER, JOSHUA

Deranian, H. Martin. "Joshua Tucker and his
portrait of the Marquis de Lafayette." AMER
DENT ASSN J 93(6) Dec '76, 1138-39 (6 ref.)

917 UMBILICAL CANCER
      Key, Jack D., David A. E. Shephard, and Walt-
      man Walters. "Sister Mary Joseph's nodule
      and its relationship to diagnosis of carci-
      noma of the umbilicus." MINN MED 59(8) Aug
      '76, 561-66 (19 ref.)
918 UNITED STATES
      Commager, Henry Steele. "10 turning points
      in our history." MOD MATURITY 19(3) Jun-Jul
      '76, 40-42
919 UNITED STATES--COLONIAL PERIOD
      Clagett, Kathleen, and Brooke Thompson-Mills.
      "The White Pine War." HORTICULTURE 54(12)
      Dec '76, 12-17
920 UNITED STATES--19th CENTURY
      "The promised land: settling the west." NAT
      PARKS & CON MAG 50(11) Nov '76, 11-13
921 UNITED STATES--MEDICAL ASPECTS--19th CENTURY
      Breeden, James O. "States-rights medicine in
      the Old South." NY ACAD MED BULL 52(3) Mar-
      Apr '76, 348-72 (50 ref.)
922 UNITED STATES--AIR FORCE
      Goodson, Wayne. "Learning to fly in the Air
      Force." AIR UNIV REV 27(5) Jul-Aug '76, 35-
      40
923 -----
      Sylva, Dave. "The Air Force story." AEROSPACE
      SAFE 32(7) Jul '76, 4-9
924 UNITED STATES--AIR FORCE--FORECASTS
      Ruhl, Robert K. "In the Tricentennial year
      2076." AIRMAN 20(7) Jul '76, 24-32
925 UNITED STATES--AIR FORCE--VETERINARY MEDICINE
      Irving, George W. "The veterinary service of

the United States Air Force: its contributions to comparative medical research." AMER
VET MED ASSN J 169(1) Jul 1, '76, 117-19
(3 ref.)

926   UNITED STATES--ARMY--19th CENTURY
Ebel, Wilfred L. "Forty miles a day on beans
and hay." SOLDIERS 31(7) Jul '76, 36-39

927   UNITED STATES--ARMY--CIVIL WAR--ORGANIZATION
Huston, James A. "Challenging the logistics
status quo during the Civil War." DEFENSE
MANAGEMENT J 12(3) Jul '76, 25-33 (6 ref.)

928   UNITED STATES--ARMY--CORPS OF ENGINEERS
Gordon, Roy. "Engineering for people, 200
years of Army public works." MIL ENG 68(443)
May-Jun '76, 180-85

929   UNITED STATES--ARMY--FINANCE
Mann, Jo Ann. "Keepin' the money comin'."
SOLDIERS 31(7) Jul '76, 17

930   UNITED STATES--ARMY--INSIGNIA
Wright, Donald C. "Bars, stars and General
Washington." SOLDIERS 31(2) Feb '76, 17

931   UNITED STATES--ARMY--ORGANIZATION
Murphy, James. "The evolution of the general
staff concept." DEFENSE MANAGEMENT J 12(3)
Jul '76, 34-39 (21 ref.)

932   UNITED STATES--ARMY--VETERINARY MEDICINE
Spertzel, Richard O. "U.S. Army veterinarians in biomedical research, from seed to
harvest." AMER VET MED ASSN J 169(1) Jul 1,
'76, 115-16

933   UNITED STATES--BICENTENNIAL CELEBRATIONS, ETC.
Gillies, Jean. "How farm people said . . .
happy birthday America." FARM J 100(11) Nov

'76, 47-49

-----. See also SAILING VESSELS 786-88

934   UNITED STATES--BICENTENNIAL CELEBRATIONS,
      ETC.--PICTORIAL WORKS
      "America gets it on." SOLDIERS 31(12) Dec.
      '76, 10-14

935   UNITED STATES--BICENTENNIAL CELEBRATIONS,
      ETC.--TELEPHONE DIRECTORIES
      Cobb, W. Montague. "A unique Bicentennial
      contribution by a corporate giant." NAT
      MED ASSN J 68(4) Jul '76, 261-62

936   UNITED STATES BICENTENNIAL CELEBRATIONS,
      ETC.--WEATHER
      Parmenter, Frances C. "From above: our Bi-
      centennial weather." WEATHERWISE 29(5) Oct
      '76, 234-35

937   UNITED STATES--BOUNDARIES
      Van Zandt, Franklin K. "Boundaries of the
      United States and the several states." US
      GEOL SURV PROF PAP (909) '76, 1-191

938   UNITED STATES--CENTENNIAL CELEBRATIONS, ETC.
      Cheney, Lynne Vincent. "1876, its artifacts
      and attitudes, returns to life at Smithson-
      ian." SMITHSONIAN 7(2) May '76, 37-49

939   -----
      Dempewolff, Richard F. "How far have we
      come since the Centennial exposition?" POP
      MECH 146(1) Jul '76, 79-81

940   -----
      "1876: a Centennial exhibition." AMER MACH
      120(7) Jul '76, 114-15

941   -----
      "1876: a Centennial exhibition." COMP AIR

MAG 81(9) Sep '76, 6-10

942 -----

Epstein, Marc J. "Centennial." SOLDIERS 31
(7) Jul '76, 33-35

943 -----

"Our ceramic heritage." AMER CERAMIC SOC BULL
55(7) Jul '76, 677-79

944 UNITED STATES--CIVIL WAR

"The Civil War, triumph of a union." NAT
PARKS & CON MAG 50(10) Oct '76, 8-10

945 -----

Davis, Daniel T. "The air role in the War
Between the States." AIR UNIV REV 27(5) Jul-
Aug '76, 13-29 (69 ref.)

946 -----

Frost, Robert W. "I like sogering first
rate." SOLDIERS 31(7) Jul '76, 25-28

947 UNITED STATES--CIVILIZATION

Freese, Arthur S. "The flowering of American
culture." MOD MATURITY 19(3) Jun-Jul '76,
47-51

948 -----

Lilienthal, David E. "New opportunities for
'underdeveloped' America to seize." SMITH-
SONIAN 7(4) Jul '76, 108-15

949 UNITED STATES--COAST GUARD--CIVIL ENGINEERING

High, Jeffrey P. "Coast Guard public works."
MIL ENG 68(443) May-Jun '76, 194-97

UNITED STATES--COAST SURVEY. See Hassler,
Rudolph 384

950 UNITED STATES--COMMERCE

"To the farthest port . . ." NAT PARKS &
CON MAG 50(8) Aug '76, 10-12

951 UNITED STATES--CONTINENTAL ARMY
        Carter, Duncan A., and Michael S. Lancaster.
        "They would endure." SOLDIERS 31(7) Jul '76,
        6-9
952 -----
        Collins, James L., Jr. "Management of the
        'mixed multitude' in the Continental Army."
        DEFENSE MANAGEMENT J 12(3) Jul '76, 7-13 (8
        ref.)
953 -----
        Silva, Eileen. "'Our want of gunpowder is in-
        conceivable'--George Washington." GUNS 22
        (6-1) Jan '76, 38-39, 70-71
954 UNITED STATES--CONTINENTAL ARMY--CAMP FOLLOWERS
        Hake, Janet. "They also served." SOLIDERS 31
        (6) Jun '76, 50-52
955 UNITED STATES--CONTINENTAL ARMY--FOOD AND NU-
        TRITION
        Manchester, Katharine E. "General Washington
        and the patriot soldiers." AMER DIETET ASSN
        J 68(5) May '76, 421-33 (32 ref.)
956 UNITED STATES--CONTINENTAL ARMY--MANAGEMENT
        Henderson, H. James. "The Continental Con-
        gress: management by Committee and Board."
        DEFENSE MANAGEMENT J 12(3) Jul '76, 3-6
957 UNITED STATES--CONTINENTAL ARMY--MEDICAL AS-
        PECTS
        Saffron, Morris H. "Confrontation in New
        Jersey." MED SOC NJ J 73(12) Dec '76, 1081-
        87
        UNITED STATES--CONTINENTAL ARMY--MEDICAL AS-
        PECTS. See also TILTON, JAMES 903

UNITED STATES--CONTINENTAL ARMY--MILITARY EN-
GINEERING. See BECHET DE ROCHE FONTAINE,
ETIENNE 88; GRIDLEY, RICHARD 371;
PORTAIL, LOUIS LE BEGUE 728; PUTNAM,
RUFUS 754

958 UNITED STATES--CONTINENTAL ARMY--PERSONAL
NARRATIVES
Nonte, George E. "A letter home, 1776."
GUNS 22(6-1) Jan '76, 45, 68-69

959 UNITED STATES--CONTINENTAL ARMY--SUPPLIES AND
STORES
"Continental Army logistics--clothing supply." ARMY LOGISTICIAN 8(2) Mar-Apr '76,
28-32

960 -----
"Continental Army logistics--engineer, or-
dinance, and medical support." ARMY LOGIS-
TICIAN 7(5) Sep-Oct '75, 24-28

961 -----
"Continental Army logistics--food supply."
ARMY LOGISTICIAN 8(1) Jan-Feb '76, 24-29

962 -----
"Continental Army logistics--the framework
and the funds." ARMY LOGISTICIAN 7(3) May-
Jun '75, 18-21

963 -----
"Continental Army logistics-an overview."
ARMY LOGISTICIAN 8(4) Jul-Aug '76, 12-13

964 -----
"Continental Army logistics--the Quartermas-
ter and Commissary departments." ARMY LOGIS-
TICIAN 7(4) Jul-Aug '75, 28-32

965 -----
     "Continental army logistics-transportation."
     ARMY LOGISTICIAN 7(6) Nov-Dec '75, 24-27
966 UNITED STATES--CONTINENTAL ARMY--VETERINARY
     MEDICINE
     Miller, Everett B. "Veterinary-farriery ser-
     vices in the Continental Army--April 1775-
     May 1777." AMER VET MED ASSN J 169(1) Jul 1,
     '76, 106-14 (66 ref.)
     UNITED STATES--CONTINENTAL NAVY. See BARNEY,
     JOSHUA 85; NAVAL HISTORY--REVOLUTIONARY WAR
     636-37
967 UNITED STATES--FORECASTS
     Macrae, Norman. "United States can keep
     growing--and lead--if it wishes." SMITHSON-
     IAN 7(4) Jul '76, 34-40
968 UNITED STATES--INTERNATIONAL RELATIONS
     Meuse, Barry M. "After the Bicentennial--
     the end of an era?" AIR UNIV REV 27(5) Jul-
     Aug '76, 2-12 (23 ref.)
969 UNITED STATES--NATIONAL OCEAN SURVEY
     Powell, Allen L., "Surveys for engineering
     and science." MIL ENG 68(443) May-Jun '76,
     198-201
970 UNITED STATES--NATIONAL SECURITY
     Pogue, Forrest C. "Economy before prepared-
     ness." DEFENSE MANAGEMENT J 12(3) Jul '76,
     14-18
971 UNITED STATES--NATIONAL PARK SERVICE
     Albright, Horace, Newton B. Drury, Conrad
     L. Wirth, and Ronald H. Walker. "Former Di-
     rectors speak out." AMER FOR 82(6) Jun '76,
     27-31, 50-51

UNITED STATES--NAVY. See MARINE ENGINEERING
527

UNITED STATES--NAVY--WORLD WAR, 1939-1945. See
NAVAL HISTORY--WORLD WAR, 1939-1945 638

972 UNITED STATES--NAVY--CIVIL ENGINEERING
Johnston, Judith. "Navy public works--1776-
1976." MIL ENG 68(443) May-Jun '76, 186-93

973 -----
Transano, Vincent A. "Bureau of Yards and
Docks, 1842-1966, NAVFAC: 1966 . . ." NAVY
CIVIL ENG 17(2) Sum '76, 4-5, 28

UNITED STATES--NAVY--CIVIL WAR. See NAVAL HIS-
TORY--CIVIL WAR 635

974 UNITED STATES--NAVY--ORGANIZATION
Hooper, Edwin B. "Developing naval concepts:
the early years." DEFENSE MANAGEMENT J 12(3)
Jul '76, 19-24 (11 ref.)

975 UNITED STATES--POLITICS AND GOVERNMENT
Commager, Henry Steele. "A generation of
heros." NAT PARKS & CON MAG 50(7) Jul '76,
9-13

976 -----
Palmer, Lane. "Can we stay united?" FARM J
100(7) Jun/Jul '76, 18-19, 40

UNITED STATES--PRESIDENTS. See LONGEVITY--
UNITED STATES--PRESIDENTS 512; VISION--
UNITED STATES--PRESIDENTS 1003

977 UNITED STATES--REVOLUTIONARY WAR
"The war for Independence, the continuing
revolution." NAT PARKS & CON MAG 50(6) Jun
'76, 10-12

978 UNITED STATES--REVOLUTIONARY WAR--BIOGRAPHY
Fondiller, Harvey V. "Men of the Revolution,

American soldiers of 1776." POP PHOTOGR 79
(1) Jul '76, 67-69, 126, 154, 216

979 -----
Kabel, Nadine B. "The gift of freedom."
AMER AGRICULTURIST 173(7) Jul '76, 30

980 UNITED STATES--REVOLUTIONARY WAR--FARMERS
"The Farmer's Revolution." NAT FUTURE FARM
24(5) Jun-Jul '76, 24-25

981 UNITED STATES--REVOLUTIONARY WAR--FRENCH PAR-
TICIPATION
Wooden, Allen C. "Dr. Jean Francois Coste
and the French Army in the American Revolution." DEL MED J 48(7) Jul '76, 397, 399-404
(16 ref.)

982 UNITED STATES--REVOLUTIONARY WAR--INTERNATION-
AL ASPECTS
Green, Timothy. "The British view of 1776
and all that: Greenwich, 1976." SMITHSONIAN
7(5) Aug '76, 64-73

983 UNITED STATES--REVOLUTIONARY WAR--MEDICAL AS-
PECTS
Cash, Philip. "The Canadian military cam-
paign of 1775-1776: medical problems and
effects of disease." AMER MED ASSN J 236(1)
Jul 5, '76, 52-56 (2 ref.)

984 UNITED STATES--REVOLUTIONARY WAR--QUOTATIONS,
MAXIMS, ETC.
"A miscellany of quotes, circa 1776." CONN
MED 40(12) Dec '76, 845-47

985 UNITED STATES--REVOLUTIONARY WAR--WEATHER
Ludlum, David M. "The weather of American
independence: the war begins." WEATHERWISE
26(4) Aug '73, 152-59 (17 ref.)

986 -----
     "Weather . . . and the Revolutionary War."
     NOAA 6(3) Jul '76, 43
987 UNITED STATES--REVOLUTIONARY WAR--WOMEN
     Hake, Janet. "Contributors to the cause,
     women who helped win the fight for indepen-
     dence." SOLDIERS 31(7) Jul '76, 15-16
988 UNITED STATES--SOCIAL CHANGE--FORECASTS
     Noble, Daniel E. "The U.S. transition from
     muscle extension to brain extension." IEEE
     PROC 64(9) Sep '76, 1418-23 (6 ref.)
989 UNITED STATES--SOCIAL CONDITIONS
     Mac Leish, Rod. "National spirit: the pendu-
     lum begins to swing." SMITHSONIAN 7(4) Jul
     '76, 28-33
990 UNITED STATES PHARMACOPEIA
     Osol, Arthur. "The Philadelphia College of
     Pharmacy and Science: its service to the
     United States Pharmacopeia and the nation."
     AMER J PHARM 148(4) Jul/Aug '76, 108-12
991 U.S.S. CONSTITUTION--MEDICAL ASPECTS
     Bradburn, H. Benjamin. "The medical log of
     Old Ironsides." CONN MED 40(12) Dec '76,
     859-68 (6 ref.)
992 UROLOGY
     Zorgniotti, Adrian W. "The creation of the
     American urologist, 1902-1912." NY ACAD MED
     BULL 52(3) Mar-Apr '76, 283-92 (32 ref.)
993 UROLOGY--18th CENTURY
     Herman, John R. "Urology at the time of the
     American Revolution." NY STATE J MED 76(8)
     Aug '76, 1362-71

994 USE STUDIES

Herner, Saul. "The library and information user-then and now." AMER SOC INFORM SCI BULL 2(8) Mar '76, 32-33

VASQUEZ DE CORONADO, FRANCISCO. See CORONADO, FRANCISCO VASQUEZ DE 208

VEGETABLE INDUSTRY AND TRADE. See FRUIT INDUSTRY AND TRADE 347-48

995 VENTILATION

Klauss, A. K., R. H. Tull, L. M. Roots, and J. R. Pfafflin. "History of the changing concepts in ventilation requirements." ASHRAE J 18(7) Jul '76, 43-44

996 VETERINARY DRUGS--THERAPEUTIC USE

Stowe, C. M. "History of veterinary pharmacotherapeutics in the United States." AMER VET MED ASSN J 169(1) Jul 1, '76, 83-89 (103 ref.)

VETERINARY MEDICINE. See LIAUTARD, ALEXANDRE 486; LIVESTOCK--VETERINARY MEDICINE 504

997 VETERINARY MEDICINE--EDUCATION

Armistead, W. W. "The ascent of veterinary medical education." AMER VET MED ASSN J 169 (1) Jul 1, '76, 38-41 (15 ref.)

998 VETERINARY MEDICINE--INTERNATIONAL ASPECTS

Maurer, Fred D. "The international influence of United States veterinary medicine." AMER VET MED ASSN J 169(1) Jul 1, '76, 70-73 (6 ref.)

999 VETERINARY MEDICINE--IOWA

Sexton, J. W. "A Bicentennial reflection on veterinary medicine in Iowa." IOWA STATE UNIV VETERINARIAN 38(3) '76, 96-103

1000 VETERINARY PUBLIC HEALTH

Steele, James H. "Veterinary public health in the United States, 1776 to 1976." AMER VET MED ASSN J 169(1) Jul 1, '76, 74-82 (19 ref.)

VIKING MISSION. See MARS(PLANET) 528-29

1001 VIOLENCE

Somers, A. R. "Health policy 1976: violence, television and American youth." ANN INTERN MED 84(6) Jun '76, 743-45 (17 ref.)

1002 VISION--FAMOUS PERSONS--COLONIAL PERIOD--PIC-TORIAL WORKS

Groffman, Sidney. "The eyes of liberty." AMER OPTOMET ASSN J 47(8) Aug '76, 1017-40 (34 ref.)

1003 VISION--UNITED STATES--PRESIDENTS

Gerber, Paul C. "The eyes of the Presidents." AMER OPTOMET ASSN J 47(10) Oct '76, 1245-60 (33 ref.)

1004 WALKER, MARY EDWARDS

Goldowsky, Seebert J. "Mary Edwards Walker, M.D., 1832-1919." RHODE ISLAND MED J 59(3) Mar '76, 118-26, 140-42 (28 ref.)

1005 WAR WOUNDS

Trueta, Joseph. "Reflections on the past and present treatment of war wounds and frac-tures." MIL MED 141(4) Apr '76, 255-58 (19 ref.)

1006 WARING, GEORGE

"George Waring: giving sanitation status." CIVIL ENG-ASCE 46(7) Jul '76, 81-82

WARREN, JOHN. See MEDICINE--18th CENTURY 567; WARREN, JOSEPH 1008

1007 WARREN, JOSEPH

Hussey, Hugh H. "The Brothers Warren, American Patriots." AMER MED ASSN J 235(24) Jun
14, '76, 2635-36, 2658 (4 ref.)

1008 -----

"Physician, politician, and patriot--1776."
ROCKY MT MED J 73(3) May-Jun '76, 135-36
(4 ref.)

-----. See also MEDICINE--18th CENTURY 567

1009 WASHING MACHINES

"The birth of the modern washer: a present
for Mrs. Blackstone." EXXON CHEM MAG 9(3) '76
12-13

1010 WASHINGTON, D.C.

Maury, William M. "Washington since Washington." MOD MATURITY 19(3) Jun-Jul '76, 52-55

1011 WASHINGTON, GEORGE

Clepper, Henry. "George Washington's trees."
AMER FOR 82(8) Aug '76, 22-25

1012 -----

"George Washington wool grower." NAT WOOL
GROW 66(2) Feb '76, 28-29

1013 -----

Weaver, Neal, "George Washington: the American Cincinnatus and America's first scientific farmer." GARDEN J 26(1) Feb '76, 8-11

-----. See also MOUNT VERNON--PICTORIAL WORKS
628

1014 WASHINGTON, GEORGE--FLOWER GARDENING

Fisher, Robert B. "Following Washington down
the garden path." LANDSCAPE ARCHIT 66(3) May
'76, 254-58

1015 WASHINGTON, GEORGE--LAST ILLNESS AND DEATH
    Scheidemandel, Heinz H. E. "Did George
    Washington die of Quinsey?" ARCH OTOLARYNGOL
    102(9) Sep '76, 519-21 (16 ref.)
1016 WASHINGTON, GEORGE--MEDICAL ASPECTS
    La Force, F. Marc. "Medical notes on the life
    of George Washington." AMER MED WOMEN ASSN J
    31(1) Jan '76, 20-21, 24-25, 29 (21 ref.)
1017 WASHINGTON, GEORGE--RELIGION
    Davis, Edwin S. "The religion of George Wash-
    ington." AIR UNIV REV 27(5) Jul-Aug '76, 30-3⁴
    (20 ref.)
1018 WATER DISTRIBUTION
    La Nier, J. Michael. "Historical development
    of municipal water systems in the United
    States, 1776-1976." AMER WATER WORKS ASSN J
    68(4) Apr '76, 173-80 (6 ref.)
    -----. See also LATROBE, BENJAMIN 479; SEWAGE
    DISPOSAL 812
1019 WATER DISTRIBUTION--FORECASTS
    Wiedeman, John H. "Distribution: what's com-
    ing down the pipe?" WATER WASTES ENG 13(7)
    Jul '76, 56-58
1020 WATER PURIFICATION
    Fielding, Liz. "A revolutionary idea--water
    for the people." WATER WASTES ENG 13(7) Jul
    '76, 16-19
1021 -----
    Storck, William J. "Wastewater's future is
    cloudy." WATER WASTES ENG 13(7) Jul '76,20-22
1022 -----
    Wolman, Abel. "200 years of water service."
    AMER WATER WORKS ASSN J 68(8) Aug '76, A13-

A18 (17 ref.)

-----. See also INFLATION (FINANCE)--WATERWORKS
430

1023 WATER PURIFICATION--FORECASTS
Heckroth, Charles W. "Reaching toward 2076."
WATER WASTES ENG 13(7) Jul '76, 26-28, 95

1024 -----
Thompson, A. Frederick, and James H. Dougherty. "AWT: the future depends on understanding." WATER WASTES ENG 13(7) Jul '76,
82, 114

1025 WATER PURIFICATION--LAWS AND REGULATIONS
Bain, R. C. "PL-92-500 vs. local priorities."
WATER WASTES ENG 13(7) Jul '76, 72-74, 101

1026 WATER SUPPLY--FORECASTS
Romano, James A. "Where will we find the
water?" WATER WASTES ENG 13(7) Jul '76, 50-
52, 54

1027 WATERWORKS
Conley, Wilber F., Jr. "What do you do when
the plant gets old?" WATER WASTES ENG 13(7)
Jul '76, 84-87

1028 WATERWORKS--BETHLEHEM, PENNSYLVANIA--18th
CENTURY
"1762 waterworks restored." WATER SEWAGE
WORKS 123(7) Jul '76, 40-41

1029 WATERWORKS--CONTROL EQUIPMENT
Norkis, Charles M., and Harold D. Gilman.
"Automation: a short history, but a long
future." WATER WASTES ENG 13(7) Jul '76,
97-98, 100

1030 WATERWORKS--EQUIPMENT
Autry, John S. "Equipment is on the move."

WATER WASTES ENG 13(7) Jul '76, 91-93

WATSON, THOMAS A. See TELEPHONE 881

1031 WATT, JAMES

Gettelman, Ken. "The spirits of '76." MOD MACH SHOP 49(2) Jul '76, 65-78

1032 WAYNE, ANTHONY

Dean, Sandy. "A spirit of '76." WEST HORSE 41(6) Jun '76, 93

WAYNE, "MAD ANTHONY." See WAYNE, ANTHONY 1032

1033 WEATHER

Rothovius, Andrew E. "Pioneers of American weather science." COMP AIR 80(12) Dec '75, 12-17; 81(1) Jan '76, 10-11

1034 WELLS

"A Bicentennial history of the water well industry." WATER WELL J 30(7) Jul '76, 45-54

1035 WEST POINT, NEW YORK-MILITARY ENGINEERING

Palmer, Dave R. "Fortress West Point: 19th century concept in an 18th century war." MIL ENG 68(443) May-Jun '76, 171-74

-----. See also OBSTACLES (MILITARY SCIENCE) 664-65

WESTERN UNION TELEGRAPH COMPANY. See TELEPHONE 882

1036 WHEAT--HARVESTING

Benne, Erwin J. "An old-fashioned wheat harvest." HOARD'S DAIRY 121(13) Jul 10, '76, 794, 802

WHITE PINE WAR. See UNITED STATES--COLONIAL PERIOD 919

1037 WHITEWARE

"Whiteware from the beginning." CER IND 107

          (1) Jul '76, 24-25, 41, 44

1038 WHITNEY, ELI
          Hirschfeld, Fritz. "The ups and downs of Eli
          Whitney." MECH ENG 98(10) Oct '76, 26-27

1039 -----
          Smith, J. J. "Eli Whitney--1765-1825." SCI
          & CHILD 13(4) Jan '76, 35-36

1040 WILDLIFE
          Gilchrist, Rick. "American wildlife in his-
          torical perspective." TENN CONSERV 42(6) Jul
          '76, 8-9

1041 -----
          Nowak, Ronald M. "Our American wildlife:
          1776-1976." NAT PARKS & CON MAG 50(11) Nov
          '76, 14-18

1042 -----
          Rockcastle, Verne N. "American wildlife--
          then and now." SCI & CHILD 13(4) Jan '76,
          9-12 (8 ref.)

1043 WILDLIFE--ENDANGERED SPECIES--UNITED STATES
          "Threatened, endangered and extinct American
          wildlife." NAT PARKS & CON MAG 50(11) Nov
          '76, 16-17

1044 WILDLIFE--MISSOURI
          Howell, Phil, "Missouri's wildlife trail,
          part II, 1937-1976." MO CONSERV 37(7) Jul
          '76, 26-61

1045 -----
          McKinley, Dan. "Missouri's wildlife trail,
          part I, 1700-1936." MO CONSERV 37(7) Jul '76,
          1-25

1046 WILDLIFE CONSERVATION
          Lovejoy, Thomas. "We must decide which

species will go forever." SMITHSONIAN 7(4)
Jul '76, 52-59

1047 WILSON, MATTHEW
Marvil, James E. "Matthew Wilson, D.D., M.D.,
1729-1790." DEL MED J 48(7) Jul '76, 391-96
(4 ref.)

1048 WINTHROP, JOHN
"John Winthrop (1714-1779)." ASTRONOMY 4(7)
Jul '76, 30

1049 -----
"John Winthrop--learned astronomer, physi-
cist." DESIGN N 31(13) Jul 4, '76, 19

1050 WINTHROP, JOHN, JR.
"John Winthrop, Jr. (1606-1676)." ASTRONOMY
4(7) Jul '76, 30

1051 WISCONSIN HEART CLUB
Madison, Frederick W. "The Wisconsin Heart
Club." WIS MED J 75(12) Dec '76, 18-20 (4
ref.)

1052 WISTAR, CASPAR
Adcock, Louis H. "Wistar, Hutchinson and
Rittenhouse, three Pennyslvania scientists."
CHEMISTRY 49(1) Jan-Feb '76, 17-18 (7 ref.)

1053 WITCHCRAFT--NEW ENGLAND--17th CENTURY
Caporael, Linnda R. "Salem's witches: was
there a Satan in the rye?" HORTICULTURE
54(6) Jun '76, 12-19

1054 WOMEN AS VETERINARIANS
Calhoun, M. Lois, and K. A. Houpt. "Women
in veterinarian medicine." CORNELL VET 66
(4) Oct '76, 455-75 (8 ref.); 67(1) Jan
'77, 1-23 (26 ref.)

WOMEN PHYSICIANS. See WALKER, MARY EDWARDS
1004

WOMEN PHYSICIANS--19th CENTURY. See ZAKRZEWSKA,
MARIE 1064

1055 WOMEN PHYSICIANS--ILLINOIS--19TH CENTURY
Johnson, Martha. "Struggle and triumph for
Illinois' first women physicians." ILL MED J
149(3) Mar '76, 291-95 (6 ref.)

1056 WOOD, ROBERT WILLIAMS
Strong, John. "Robert Williams Wood." APP
OPTICS 15(7) Jul '76, 1741-43 (8 ref.)

1057 WOODHOUSE, JAMES
Beer, John J. "An answer to Dr. Joseph
Priestley's considerations on the doctrine
of phlogiston." J CHEM ED 53(7) Jul '76,
414-18

1058 WOODWORKING INDUSTRIES
"200 years of woodworking." WOOD WOOD PROD
81(6) Jun '76, 21-35

1059 WORLD POLITICS
Augelli, John P. "Evolution of America's
world posture: perception and reality."
J GEOGRAPH 75(1) Jan '76, 7-27 (22 ref.)

1060 YELLOW FEVER
"American doctors lead victory over Yellow
Fever." MED ASSN STATE ALA J 45(12) Jun '76,
24

1061 -----
Cates, Gerald L. "The St. Mary's Yellow Fever
epidemic of 1808: Georgia's first confron-
tation." MED ASSN GEORGIA J 65(7) Jul '76,
287-90 (7 ref.)

1062 -----

"Update on Bilious Plague-Pennyslvania,
1793." MORBID MORTAL WEEK REP 25(25 pt. 2)
Jul 2, '76, 6 (3 ref.)

1063 YORKTOWN, VIRGINIA--SIEGE, 1781--MILITARY EN-
GINEERING

Haskett, James N. "Military engineers at
Yorktown, 1781." MIL ENG 68(443) May-Jun
'76, 175-79

YOUNGSTOWN, OHIO--MEDICAL CARE. See DUTTON,
CHARLES 241

1064 ZAKRZEWSKA, MARIE

Greenwood, Ronald D. "Marie Zakrzewska, M.D.,
notable physician." NY STATE J MED 76(8)
Aug '76, 1339-41 (2 ref.)

1065 ZOOS--VETERINARY MEDICINE

Schroeder, Charles R. "Zoo medicine: yesterday and today." AMER VET MED ASSN J 169(1)
Jul 1, '76, 61-69

Journal Title Index

AIA J. see American Institute of Architects. Journal
ASHRAE Journal 18(7) Jul '76; Bicentennial issue:17, 135, 339, 416, 766, 985
ASTM Standardization News 4(7) Jul '76:494
Aerospace Safety 32(7) Jul '76:923
Agricultural Engineering 57(12) Dec '76:5
Agricultural History 50(1) Jan '76; Bicentennial symposium-- two hundred years of American agriculture:7
Air University Review 27(5) Jul-Aug '76:922, 945, 968, 1017
Airman 20(7) Jul'76:924
Alabama Journal of Medical Science 13(3) Jul '76:505
Alcohol Health and Research World Sum '76:20
American Agriculturist 173(3,7) Mar, Jul '76:294, 979
American Aviation Historical Society. Journal 21(1-2) Spr, Sum '76:19, 81
American Bee Journal 116(2-7) Feb-Jul '76:92, 94-95
American Ceramic Society. Bulletin 55(1, 7) Jan, Jul '76:141-142, 943
American Congress on Surveying and Mapping. Bulletin (52, 54) Feb, Aug '76:373, 513
American Dental Association. Journal 93(1, 6) Jul, Dec '76: 223, 271, 465, 916
American Dietetic Association. Journal 68(1, 3, 5) Jan, Mar, May and 69(1-3, 5) Jul, Aug, Sep, Nov '76: 202, 205, 231, 289, 311, 663, 955
American Druggist 174(1) Jul '76:239
American Dyestuff Reporter 65(3) Mar '76:421
American Fabrics and Fashions (106) Win/Spr '76 and (107) Spr '76; Bicentennial of American textiles 899
American Forests 82(1-12) Jan-Dec '76: 305, 322-25, 474, 522, 678, 750, 838, 912, 971, 1011
American Fruit Grower 96(7) Jul '76:348
American Gas Association Monthly 57(7-12) Jul-Dec '75 and 58(1-8) Jan-Aug '76:356
American Geophysical Union. Transactions 57(7) Jul '76:63, 342
American Geriatrics Society. Journal 24(10) Oct '76:343

Annals of Science 33(1) Jan '76:1, 535
Annals of Surgery 184(5) Nov '76:803
Annual Review of Microbiology 30, '76:191
Applied Optics 15(7) Jul '76:23, 232, 237, 631, 784, 801,
  836, 1056
Archives of Dermatology 112(Special Issue) Nov 29, '76
  Special Centennial edition:29, 131, 224-225, 417, 464
Archives of Otolaryngology 102(9) Sep '76:481, 1015
Archives of Surgery 111(1) Jan '76:234
Arizona Highways 52(1, 7) Jan, Jul '76:24, 350, 420
Arizona Medicine 33(1-2,6) Jan, Feb, Jun '76:545, 560
Army Logistician 7(3-6) May-Jun, Jul-Aug, Sep-Oct, Nov-Dec,
  '75 and 8(1-2, 4) Jan-Feb, Mar-Apr, Jul-Aug '76:959-65
Astronomy 4(7) Jul '76:47-48, 51, 53-56, 59-60, 83, 113, 181,
  190, 211, 374, 407, 478, 606, 775, 783, 814, 1048, 1050
Automotive Engineering 84(7) Jul '76:449-50
Automotive Industries 155(1) Jul 1, '76; The great American
  automotive story:67-70, 73-76, 532, 626-27
Avian Diseases 20(4) Oct-Dec '76:79

Bell System Technical Journal 55(3) Mar '76:869, 879
Bioscience 26(5, 12) May, Dec '76:87, 459
Brick and Clay Record 169(1) Jul '76: Special Bicentennial
  issue:4, 116-17, 123, 718, 765
British Medical Journal 1(6025) 26 Jun '76 and 2(6026) 3 Jul
  '76:710
Bulletin of the History of Medicine 50(2-4) Sum, Fal, Win '76:
  144, 189, 702, 730, 829-30, 892

Canadian Journal of Surgery 19(5) Sep '76:570
Cancer Research 36(7 pt. 1) Jul '76:781
Cattleman 63(1) Jun '76:393
Ceramic Industry 107(1) Jul '76; CI's Bicentennial issue:139-
  40, 270, 367, 1037
Ceramics Monthly 24(5) May '76:143
Cereal Foods World 21(7) Jul '76:369, 719
Chemical and Engineering News 54(15) Apr 6, '76; Centennial,
  American Chemical Society, 1876-1976; (36) Aug 30, '76:28,
  102, 148, 154, 156-57, 160-64
Chemical Week 118(7) Feb 18, '76:147
Chemistry 48(8-11) Sep-Dec '75 and 49(1, 2, 6) Jan-Feb, Mar,
  Jul-Aug '76:146, 149, 151, 153, 458, 777, 821, 1052
Chemistry in Britain 12(12) Dec '76:158-59
Chemtech 6(2) Feb '76:375
Chilton's Oil and Gas Energy 2(1) Jan '76:273
Civil Engineering-ASCE 46(7) Jul '76; Special issue, U.S.
  Bicentennial:37, 119, 218, 269, 473, 479, 697, 824, 907,
  1006
Cleft Palate Journal 13(4) Oct '76:182

Industry Week 190(1) Jul 5, '76:428, 475
Institutions/Volume Feeding 79(1) Jul 1, '76:320
Interdisciplinary Science Reviews 1(2) Jun '76:805
Iowa State University Veterinarian 38(3) '76:999
Iron Age 217(1, 7, 15) Jan 5, Feb 16, Apr 12 and 218(9) Aug
   30, '76: Apr 12 issue is the Bicentennial issue:496, 518,
   520, 593
Iron and Steel Engineer 53(7) Jul '76:843

JOGN (Journal of Obstetric, Gynecologic and Neonatal Nursing)
   5(2-3) Mar-Apr, May-Jun '76:166, 732
Journal of Chemical Education 53(7-8, 12) Jul, Aug, Dec '76:
   104, 109, 145, 150, 152, 155, 203, 695, 735, 1057
Journal of Dermatologic Surgery 2(5) Nov '76:715
Journal of Geography 75(1) Jan '76; Bicentennial Issue:177,
   359, 410, 1059
Journal of Hospital Pharmacy 33(8) Aug '76:900
Journal of Medical Education 51(12) Dec '76:555
Journal of Occupational Medicine 18(12) Dec '76:426
Journal of Pediatrics 89(1) Jul '76:687
Journal of Professional Activities. (ASCE Proceedings) 101
   (EI3) Jul '75:120
Journal of School Health 46(1-2) Jan, Feb '76:539
Journal of the History of Medicine and Allied Sciences 31(3)
   Jul '76; Bicentennial issue:233, 297, 455, 572, 581, 630
Journal of Trauma 16(4) Apr '76:264

Kansas Medical Society. Journal 77(7) Jul '76:39, 526

Landscape Architecture 66(3) May '76:1014
Library of Congress. Quarterly Journal 34(2) Apr '77:101
Living Museum 38(6) Nov-Dec '76:757
Living Wilderness 40(133) Apr-Jun '76:198
Louisiana State Medical Society. Journal 128(11) Nov '76:477

Mc Call's 103(7) Apr '76; 100th Anniversary issue:515
Machine Design 48(15) Jun 24, '76:529
Man/Society/Technology 35(5) Feb '76:422-24
Marine Engineering/Log 31(8-9) Jul, Aug '76:787, 818
Marine Technology Society. Journal 10(6) Jul-Aug '76:668, 845
Maryland State Medical Journal 25(1, 5, 7, 10) Jan, May, Jul,
   Oct '76:362, 531, 677, 731
Materials Evaluation 34(2, 12) Feb, Dec '76:493, 651
Mathematics Teacher 69(1-8) Jan-May, Oct-Dec '76:22, 84, 242,
   244, 355, 514, 533-34
Mechanical Engineering 98(1, 5-12) Jan, May-Dec, '76:106
   276, 537, 769, 810, 840, 850, 894, 1038
Medical Association of Georgia. Journal 65(7) Jul '76:378,
   506, 1061

Medical Association of the State of Alabama. Journal 45(11-12)
    May, Jun '76:38, 389, 709, 822, 901, 1060
Medical Society of New Jersey. Journal 73(1, 3-5, 9, 11-12)
    Jan, Mar, Apr, May, Sep, Nov, Dec '76:38, 126, 171, 188,
    263, 284, 326, 400, 429, 507, 525, 542, 576, 596, 648, 795,
    957
Medical Times 104(7) Jul '76:579
Medical World News 16(27) Dec 15, '75 and 17(14, 23) Jun 28,
    Oct 25, '76:563-64, 745
Meteoritics 10(3) Sep '75:594
Metropolitan Life Insurance Company, New York. Statistical
    Bulletin 57 Feb, Mar, Apr, Jul-Aug, Nov'76 and 58 Jan '77:
    267, 509-12, 727
Michigan Academician 8(4) Spr '76:364
Military Engineering 68(441-44, 446) Jan-Feb, Mar-Apr, May-
    Jun, Jul-Aug, Nov-Dec '76:44, 88, 124, 201, 275, 278-79,
    371, 598, 665, 728, 754, 928, 949, 969, 972, 1035, 1063
Military Medicine 141(4, 11) Apr, Nov '76:368, 1005
Mining Congress Journal 62(2) Feb '76:602
Mining Engineering 28(7) Jul '76:603
Minnesota Medicine 59(7-8, 12) Jul, Aug, Nov '76:327, 388, 917
Missouri Botanical Garden. Bulletin 64(6) Jun '76:460
Missouri Conservationist 37(7) Jul '76; Special issue, Mis-
    souri's wildlife trail; 1700-1976:1044-45
Modern Castings 66(1-8, 10-12) Jan-Aug, Oct-Dec '76:330-32
Modern Concepts in Cardiovascular Disease 45(1) Jan '76:134
Modern Healthcare 6(1) Jul '76:396, 540
Modern Machine Shop 49(2) Jul '76:1031
Modern Maturity 19(1, 3) Feb-Mar, Jun-Jul '76:222, 413, 641,
    839, 918, 947, 1010
Modern Paint and Coatings 66(7) Jul '76:680
Morbidity and Mortality Weekly Report 25(25 pt. 2) Jul 2, '76;
    Bicentennial issue: 235, 243, 401, 523, 536, 797, 828, 860,
    915, 1062
Morton Arboretum Quarterly 12(1-3) Spr, Sum, Aut '76:111, 530
Mosaic 7(4) Jul-Aug '76:246
Motor Age 95(6) Jun '76; The Great American automotive story
    in a special Bicentennial issue:64-65, 71-72, 77-78, 904

NOAA 6(1, 3-4) Jan, Jul, Oct '76:384, 786, 788, 986
National Future Farmer 24(5) Jun-Jul '76:293, 980
National Medical Association. Journal 68(4) Jul '76:935
National Parks and Conservation Magazine 50(1-12) Jan-Dec,'76:
    25, 208, 221, 266, 456, 484, 644, 920, 944, 950, 975, 977,
    1041, 1043
National Wildlife 14(1, 6) Dec-Jan, Oct-Nov '76 and 15(1)
    Dec-Jan '77:287, 485, 913
National Wool Grower 66(2, 4, 6-7, 11) Feb, Apr, Jun, Jul,
    Nov '76:97, 467, 592, 815, 817, 1012

Professional Safety 21(1-12) Jan-Dec '76:2, 785, 848
Progressive Architecture 57(7) Jul '76:43
Progressive Farmer 91(7) Jul '76:212
Public Works 107(7) Jul '76:412

Quality Progress 9(7) Jul '76:755-756, 841, 897-98

RN 39(1-12) Jan-Dec '76:132, 214, 363, 427, 654-58, 666, 673,
 685
Railway Age 177(12) Jul 4, '76:759-63
Review of Surgery 33(3) May-Jun '76:387
Rhode Island Medical Journal 59(2-3, 5, 7) Feb, Mar, May, Jul
 '76:167, 349, 544, 580, 729, 770, 853, 1004
Rocks and Minerals 51(5, 7) Jun, Sep '76:604-5, 842
Rocky Mountain Medical Journal 73(3) May-Jun '76:1008
Royal Society of Medicine. Proceedings 69(6) Jun '76:642

SMPTE Journal 85(7) Jul '76; Anniversary issue:608-20
 699, 833-34, 890
Saturday Review 3(6) Dec 13, '75; America's Impact on the
 World, 1776-1976:864
Science 194(4267) Nov 19, '76:806
Science and Children 13(4) Jan '76:52, 82, 306, 340, 490, 565,
 733, 1039, 1042
Science Teacher 43(5) May '76:98
The Sciences 16(2) Mar-Apr '76:58
Scientific American 235(1, 3) Jul, Sep '76:8, 128
Sea Technology 17(5) May '76:107
Sheep Breeder and Sheepman 96(5, 7) May, Jul '76:463, 816
Sky and Telescope 51(1-2, 6) Jan, Feb, Jun '76 and 52(1) Jul
 '76:49-50, 57
Smithsonian 7(2-6) May-Aug '76:344, 468, 524, 720, 726, 808-
 9, 865, 874, 938, 948, 967, 982, 989, 1046
Soil Conservation Society of America. Proceedings of the 31st
 Meeting: Critical Conservation Choices: A Bicentennial Look.
 SCSA, Ankeny, IA, 1976:411
Soldiers 31(2,5-7, 12) Feb, May, Jun, Jul, Dec '76:370, 664,
 926, 929-30, 934, 942, 946, 951, 954, 987
South Carolina Medical Association. Journal 72(1, 5) Jan, May
 '76:3, 556
Southern Medical Journal 69(3) Mar '76:466
Surgery, Gynecology and Obstetrics 143(5) Nov '76:861
Surveying and Mapping 36(2) Jun '76; Bicentennial issue:262,
 358, 857

Tappi 59(7, 12) Jul, Dec '76:682
Telephony 190(1, 4, 21, 23) Jan 5, Jan 26, May 24, Jun 7, '76
 and 191(1, 14, 21) Jul 5, Oct 4, Nov 22, '76:875-78, 883-85,
 887-88

Tennessee Academy of Science. Journal 51(1) Jan '76:122
Tennessee Conservationist 42(6) Jul '76:303, 859, 1040
Textile World 126(1) Jan '76:895
Tobacco Reporter 103(7) Jul '76:591
Tooling and Production 42(4) Jul '76:521
Traffic Engineering 46(7, 9, 11) Jul, Sep, Nov '76:905
Trains 36(9) Jul '76:758
Translog 7(3, 7, 10) Mar, Jul, Nov '76:105, 328, 681

United Fresh Fruit and Vegetable Assn. Annual 1976:347
U.S. Bureau of Mines. Bulletin (667) '75, Mineral Facts and
  Problems:601
U.S. Geological Survey. Bulletin (1476) '76:361
U.S. Geological Survey. Professional Paper (909) '76:937
U.S. Geological Survey. Water Supply Paper (2038) '76:288
Utah Science 37(3) Sep '76:660

Virginia Medical Monthly 103(7, 10) Jul, Oct '76:773, 837

Wallaces' Farmer 101(12) Jun 26, '76:197
Water & Sewage Works 123(7) Jul '76:1028
Water & Wastes Engineering 13(7) Jul '76:108, 430, 791-94,
  826, 1019-21, 1023-27, 1029-30
Water Pollution Control Federation. Journal 48(7) Jul '76:812
Water Well Journal 30(7) Jul '76:1034
Weatherwise 26(4) Aug '73; 27(4) Aug '74; 28(3,4) Jun, Aug
  '75; 29(2,5,6) Apr, Oct, Dec '76:110, 125, 333, 508, 645,
  914, 936, 985
Welding Journal 55(7) Jul '76:255
Western Horseman 41(1-2, 6, 9-10) Jan-Feb, Jun, Sep-Oct '76:
  118, 419, 723, 1032
Western Journal of Medicine 125(1, Jul, Dec '76:129, 346, 676
  782
Wisconsin Medical Journal 75(4, 7-12) Apr, Jul-Dec '76:372,
  399, 431, 582, 849, 854, 1051
Wood and Wood Products 81(6) Jun '76:1058

Author Index

Proper Name Index

Addendum

The following items are not serial in nature but do pertain to various aspects of the bicentennial. They were uncovered in the course of the preparation of the bibliography. No claim for comprehensiveness is made (especially for government publications). The list is included solely for informational purposes.

Advances in American Medicine, Essays at the Bicentennial. John Z. Bowers, ed., New York, Josiah Macy, Jr. Foundation, 1976, 918 p.

American Public Works Association. History of Public Works in the United States, 1776–1976, Ellis L. Armstrong, ed., Chicago, The Association, 1976, 736 p.

The American Revolution. (U.S. Government Printing Office, Subject Bibliography No. 144) Washington, D.C., U.S. Government Printing Office, 1978, 14 p.

A list of some of the government documents published as a result of the bicentennial celebration. Items appearing the addendum supplement this list.

American Revolution Bicentennial Administration. The Bicentennial of the United States of America: A Final Report to the People, Washington, D.C.,

U.S. Government Printing Office, 1977, 5 volumes.

America's Highways--1776-1976. Washington, D.C.,U.S.
Government Printing Office, 1976, 553 p.

The Bicentennial Tribute to American Mathematics.
Dalton Tarwater, ed., Washington, D.C. Mathematical Association of America, 1977, 225 p.

Bordley, James, III and A. McGehee Harvey. Two Centuries of American Medicine, 1776-1976, Philadelphia, Saunders, 1976, 844 p.

Cazier, Lola. Surveys and Surveyors of the Public
Domain, 1785-1975, Washington, D.C., U.S., Government Printing Office, 1976, 228 p.

Changing Scenes in Natural Sciences, 1776-1976.
Clyde E. Goulden, ed., Philadelphia, Academy of
Natural Sciences, 1977, 362 p.

Darby, William J. Nutrition Science: An Overview of
American Genius (W.O. Atwater Memorial Lecture)
Washington, D.C., U.S. Government Printing Office,
1976, 39 p.

Federal Reserve Bank of Chicago. A Bicentennial
Chronology of Economic and Financial Events in
their Social and Political Environment, Chicago,
The Bank, 1976, 92 p.

The Importance of Minerals, 1776-1976. Washington,
D.C., U.S. Government Printing Office, 1976, 39 p.

National Academy of Sciences. Science--An American
Bicentennial View, Commentaries from a Series of
Academy Forums, Washington, D.C., National Acad-
emy of Sciences, 1977, 108 p.

On the History of Statistics and Probability. D.B.
Owen, ed., (Statistics, Textbooks and Monographs,
V. 17) New York, Dekker, 1976, 466 p.

Physician Signers of the Declaration of Independence.
    George E. Gifford, Jr., ed., New York, Science
    History, 1976, 163 p.

Royal College of Physicians of London. Medicine in
    America, 1680-1820, London, The College, 1976
    244 p.

St. Clair, Hillary W. Mineral Industry in Early
    America, Washington, D.C., U.S. Government Print-
    ing Office, 1977, 62 p.

200 Years of American Worklife. Washington, D.C.,
    U.S. Government Printing Office, 1977, 192 p.

Two Hundred Years of Flight in America: A Bicentenn-
    ial Survey, Eugene M. Emme, ed., American Astro-
    nautical Society History Series - 1) San Diego,
    Univelt, 1977, 310 p.

UCLA Conference on American Folk Medicine, 1973.
    American Folk Medicine (Publication of the UCLA
    Center for the Study of Comparative Folklore and
    Mythology - 4) Berkeley, Univ. of California
    Press, 1976, 347 p.

The University and Medicine. John Z. Bowers, ed.,
    New York, Josiah Macy, Jr. Foundation, 1977,
    239 p.

Vagtborg, Harold. Research and American Industrial
    Development, New York, Pergamon, 1976, 474 p.

Weart, Spencer R., ed., Selected Papers of Great
    American Physicists, New York, American Physical
    Society, 1976, 176 p.